Basic Flight Mechanics

Ashish Tewari

Basic Flight Mechanics

A Simple Approach Without Equations

Springer

Ashish Tewari
Department of Aerospace Engineering
Indian Institute of Technology
Kanpur, Uttar Pradesh, India

ISBN 978-3-319-80706-5 ISBN 978-3-319-30022-1 (eBook)
DOI 10.1007/978-3-319-30022-1

Softcover reprint of the hardcover 1st edition 2016

Printed on acid-free paper

This Springer imprint is published by Springer Nature
The registered company is Springer International Publishing AG Switzerland

Actioni contrariam semper et æqualem esse reactionem.—Isaac Newton

Preface

This book addresses an important gap in the aerospace literature by presenting flight mechanics in an accessible manner. It covers the flight mechanics of aircraft, spacecraft, and rockets in simple terms and purely on physical principles. Those readers who have no training in engineering mathematics, yet would like to understand how the various flight vehicles operate, will find this book especially useful.

The book contains neither any mathematical equations, data tables, nor references to technical literature. The focus is on explaining the important physical principles through simple examples and illustrations.

When the Wright brothers designed, tested, built, and flew the first heavier-than-air powered machine more than a century ago, they had little knowledge of the mathematical concepts such as the Laplace equation, integral calculus, vector algebra, and so on. But they did have the proper understanding of the physical principles behind flight mechanics, which they successfully put into practice. This book is written mainly to help a reader without an engineering education to grasp the fundamental aspects of airplane, rocket, and space flight. It is not necessary to take an engineering course or to read an engineering textbook for understanding the basic flight concepts. However, it is hoped that this book can serve as a "primer" in motivating a reader to study flight mechanics as a formal subject. For a formal mathematical and computational coverage of most of the topics covered here, the reader can refer to the author's textbook (*Atmospheric and Space Flight Dynamics – Modeling and Simulation*, Birkhaüser, Boston, 2007).

I am grateful to the editorial and production staff at Springer for their cooperation. I would also like to thank my family for their patience during the writing of this book.

Kanpur, India
December 2015

Ashish Tewari

Contents

1 Introduction 1
1.1 Classification of Flight 1
1.2 Basic Definitions 2
1.3 Archimedes Principle and Balloon Flight 6
1.4 Aerodynamic Flight 7
1.5 Basic Laws of Motion 8
1.6 Translation in a Curve 10
1.7 Coriolis Acceleration 12
1.8 Rigid-Body Dynamics 15
1.9 Flight Stability and Control 20

2 Aerodynamics 23
2.1 Relative Air Flow 23
2.1.1 Compressibility 24
2.2 Lift and Drag 25
2.2.1 Viscous Flow 27
2.2.2 Streamlined Shapes 28
2.3 How Is Lift Created? 30
2.3.1 Downwash 31
2.3.2 Pitching Moment 32
2.3.3 Pressure Distribution 33
2.4 Vorticity and Circulation 35
2.4.1 Effects of Angle-of-Attack 37
2.4.2 Finite-Wing Effect 38

3 Flight of Airplanes and Gliders: Vertical Plane 43
3.1 Introduction 43
3.2 Straight and Level Flight 44
3.2.1 Flight Envelope 46
3.3 Cruising Flight 49
3.4 Climbing Flight 52
3.5 Descending and Gliding Flight 53

3.6 Vertical Maneuvers 55
3.7 Take-Off and Landing 57
3.8 Degrees-of-Freedom and Control Surfaces 60
3.9 Speed and Altitude Control 62
3.10 Static Longitudinal Stability 65
3.11 Longitudinal Control 69
3.12 Dynamic Longitudinal Stability 70
3.13 Stalling Characteristics 72

4 Flight of Airplanes and Gliders: Horizontal Plane 73
4.1 Introduction 73
4.2 Level and Coordinated Turn 74
4.3 Lateral–Directional Dynamics 77
4.4 Static Lateral–Directional Stability 78
4.5 Dynamic Lateral–Directional Stability 80
4.6 Lateral–Directional Control 82
4.7 Spin and Recovery 85

5 Flapping and Rotary Wing Flight 87
5.1 Introduction 87
5.2 Why We Can't Fly Like the Birds 88
5.2.1 Flapping Flight Basics 88
5.2.2 Basic Flapping Mechanism 90
5.3 Rotary Wing Flight 92
5.4 Vertical Ascent, Descent, and Hover 94
5.5 The Conundrum of Forward Flight 94
5.6 Need for a Tail Rotor 97
5.7 Controls for Rotorcraft Flight 97

6 Space Flight 99
6.1 Introduction 99
6.2 Orbits 101
6.3 Orbital Maneuvers 104
6.3.1 Rocket Powered Maneuvers 104
6.3.2 Gravity-Assist Maneuver 105
6.3.3 Aero-Assisted Orbital Transfer 106
6.4 Orbital Perturbations 108
6.5 Geosynchronous and Sun-Synchronous Orbits 111
6.6 Atmospheric Entry Vehicles 112
6.6.1 Hypersonic Aerothermal Load 112
6.6.2 Ballistic and Lifting Trajectories 115
6.7 Attitude Stability Control of Spacecraft 117
6.7.1 Spin Stabilization 117
6.7.2 Gravity-Gradient Stabilization 119
6.7.3 Attitude Control 119

7 Rocket Flight 123
7.1 Introduction 123
7.2 Launch Vehicle Staging 125
7.3 Gravity-Turn Trajectory 125
7.4 Attitude Stability and Control 127

Appendix: Standard Atmosphere 133

Chapter 1
Introduction

1.1 Classification of Flight

This is a book about flight and its various basic mechanisms. In common usage, the term "flight" can apply to almost anything that moves through the air or the space, but without any support from the ground. This includes throwing a stone in the air, launching a hot-air-balloon in the sky, soaring of the birds and the airplanes, the lift-off of a rocket, and the satellites orbiting the earth. However, each one of these examples is different in the governing mode of flight. While a stone cannot stay aloft for any significant length of time, the birds, airplanes, balloons, and rockets can keep flying for hours (the satellites, for years and decades). But the flight duration can be misleading in distinguishing between the basic mechanisms involved in flight. For example, a stone projectile and a satellite are both governed by gravity, and therefore have the same mode of flight. On the other hand, the flight of airplanes and birds is governed by aerodynamic forces, that of the balloon by buoyancy, whereas a rocket is driven by the thrust of its engines. Hence, a classification of flight modes can be carried out according to the governing (or predominating) forces.

The most primary classification of flight is by the *altitude*, which is defined as the geometric height of the vehicle above the planet's surface. Since the atmospheric properties govern the nature of flight, this is a natural way of distinguishing the various flight objects. A flight taking place 100 km (or more) above the earth's surface is generally regarded as the space flight, because there is a lack of sensible atmosphere above that altitude. A flight occurring at a significantly lower altitude is termed the atmospheric flight, and a vehicle designed to operate in the air for its primary mission is called an *aircraft*. Similarly, spacecraft are the vehicles designed to primarily operate in the space. For example, all aircraft fly at altitudes less than 25 km above the earth, while most of the spacecraft do not descend below the altitude of 250 km. Hence, we have a tenfold separation between the maximum and the minimum altitudes of the aircraft and the spacecraft around the earth. The atmospheric entry vehicles such as the *Space Shuttle* and the *Soyuz* capsule are considered to be spacecraft, because their primary mission is performed in the space.

A. Tewari, *Basic Flight Mechanics*, DOI 10.1007/978-3-319-30022-1_1

Rockets are capable of operating both inside and outside the atmosphere, therefore they are considered to be in a separate category, different from both aircraft and spacecraft.

Aircraft require a support from the atmosphere called the *lift* in order to sustain themselves in flight. A further classification based upon the mechanism by which the lift is produced is thus natural. Aircraft are broadly separated into lighter-than-air (LTA) and heavier-than-air (or aerodynamic lifting) vehicles, depending, respectively, upon whether the lift is produced by buoyancy, or by the motion of the vehicle (or a part of the vehicle) through the air. Into the LTA category fall the balloons and airships, while the aerodynamic lifting vehicles are classified as being either fixed-wing, or rotary-wing craft depending upon whether their wings are fixed or rotary relative to the main structure. The fixed-wing aircraft include the airplanes and the gliders, whereas the rotary-wing vehicles (also termed the rotorcraft) are the helicopters and autogyros. Although there is no flapping-wing flight vehicle currently in operation, they can be thought of as being aerodynamic vehicles in a class separate from the fixed-wing and the rotary-wing vehicles.

Spacecraft are classified according to their mission. A satellite operating in a near orbit around the earth (altitudes less than about 2000 km) is a *low-earth orbit* (LEO) spacecraft. The orbital period of LEO spacecraft ranges from 85 to 130 min. Medium earth orbit (MEO) spacecraft, such as the global positioning system (GPS) satellites, orbit at much higher altitudes, and have orbital periods ranging between 8 and 15 h. The geosynchronous (GEO) satellites used for telecommunications purposes have their orbits in the equatorial plane, with the orbital period matching that of the earth's rotation about its axis. A spacecraft sent to the Moon is termed a lunar spacecraft, while that sent to another planet is called an interplanetary spacecraft.

Classification of the rockets is by their range and payload. They include satellite launch vehicles from small (less that 100 kg) to large (greater than 1000 kg) payloads to LEO, MEO, or GEO orbits, as well as short, medium, and long range ballistic missiles. Missiles are also classified by their launch and target positions, such as air-to-air, surface-to-air, or surface-to-surface missiles.

In the rest of this chapter, we will broadly discuss the physical concepts that enable flight.

1.2 Basic Definitions

Let us first introduce the required terminology. Three mutually perpendicular straight lines can be considered to constitute a *reference coordinate frame* from which the motion of an object can be observed. The perpendicular lines of a reference frame are called its *axes*, whereas the point at which the axes meet is called the *origin* of the reference frame. An example of a reference frame is a corner of a hall in which a flight experiment is taking place. The origin of a reference frame could be either fixed, or moving, and its axes could be either fixed in direction, or rotating about the origin while staying perpendicular to each other.

A *particle* is an infinitesimal part of matter without any spatial dimensions. It is also referred to as a point mass. The *position* of a moving particle refers to its location relative to a reference frame. The position is determined both by the distance of the particle from the origin measured along a straight line and the angles made by the same straight line with the axes of the frame. The time rate of change of position of a point is called its *velocity*, whereas the time rate of change of the velocity is the *acceleration*. If the reference frame used to measure the position, velocity, and acceleration of a particle is itself undergoing motion, then it would not produce the same measurements as those carried out in a stationary reference frame. Often for flight dynamic description, the measurements are performed in a special reference frame, whose axes do not rotate, and whose origin moves in a straight line at a constant speed. Such a reference frame is called an *inertial frame*. A *stationary frame* is an example of the inertial frame in which the origin is at rest.

The *mass* of a body refers to a property called *inertia* by virtue of which it resists any change from the state of uniform (i.e., constant speed) motion in a straight line. The state of rest is a special case of such a motion in which the speed is always zero. We can appreciate the significance of inertia when we try to push a car which has failed to start. The *force* is a general mechanism by which two given objects can interact with one another. The *gravity* is a special force by which every object attracts every other object by virtue of its mass. Newton discovered that the force of gravity between two objects is directly proportional to the product of their masses, and inversely proportional to the square of the distance between them. Hence, a massive object such as a planet attracts every object on its surface by a force of gravity, which rapidly diminishes as the objects mover further away. The Moon, although very far away, is bound by the earth's gravity in the same manner as a house or a car on the earth's surface. If we divide the earth's gravitational force by the mass of the object being pulled by it, we get the acceleration due to gravity. Two objects of different masses at the same distance from the earth's center experience the same acceleration due to gravity, as famously demonstrated by Galileo in an experiment at the leaning tower of Pisa. We will return to the concepts of inertia (mass), force, and acceleration later in this chapter.

The basic quantities needed to describe the motion of a particle—namely, the velocity, acceleration, and force—have both a magnitude (i.e., size) and a direction. Such quantities are called *vectors*, and have mathematical properties different from *scalar* quantities, such as money, temperature, time, and mass, which can be added, subtracted, multiplied, and divided arithmetically.

The magnitude of a vector indicates its length (or size) and is a scalar quantity. The magnitude of the velocity vector is termed the *speed*.

The addition of two vectors is carried out by constructing a triangle in which the magnitude of each vector is represented by the length of the straight line forming one of the two sides of the triangle. The direction of each vector is shown by an arrow, and the two vectors to be added are joined nose to tail (see Fig. 1.1a).

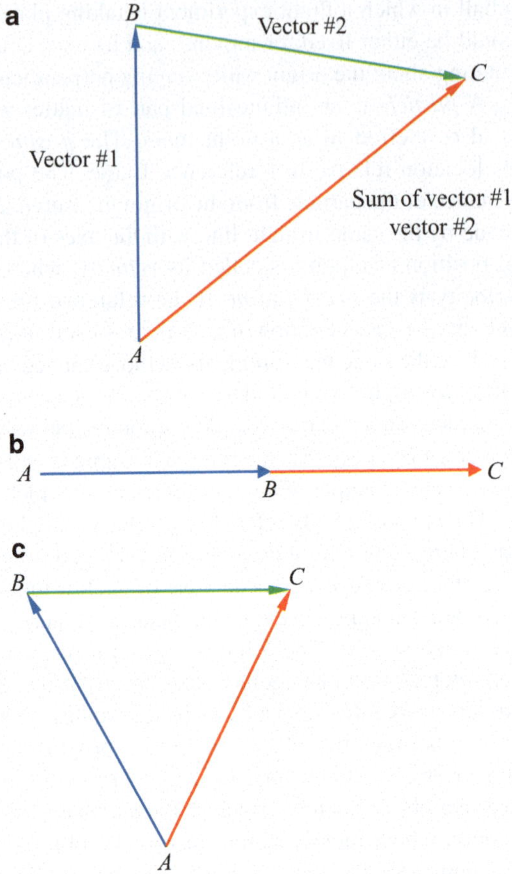

Fig. 1.1 Addition of two vectors. (**a**) A general addition of vectors *AB* and *BC* to produce the sum *AC*. The three vectors have different magnitudes and directions. (**b**) Addition of two collinear vectors, in which there is no change in the direction. (**c**) Addition of two vectors where the initial vector, *AB*, and the resultant sum, *AC*, have equal lengths. This is the special case in which the vector *AB* has only undergone a change in its direction, but not in its magnitude

The third side, *AC*, of the resulting triangle then gives the vector sum of the two vectors: *AB* and *BC*. If there is neither a change in the magnitude or the direction of a vector quantity with the time, then it is said to be unchanged, or *constant* with time.

The force applied at a distance from a given point *O* is called the *moment* of the force about *O*. The magnitude of the moment is the product of the magnitude of the force and the perpendicular distance—called the *moment arm*—measured at a 90° angle from the line of the force's application. The moment is a vector, and acts along the axis passing through *O*, and normal (i.e., at a 90° angle) to both the applied force and the moment arm. The moment is thus produced in a direction normal to the plane formed by the force and the arm. The larger the moment arm, the greater the moment produced for the same force. The direction of the moment vector is given by the *right-hand rule*, such that the fingers of the right-hand curve from the direction of the arm toward the applied force, and the extended thumb indicates the moment vector. Before applying the right-hand rule, the force must be transferred

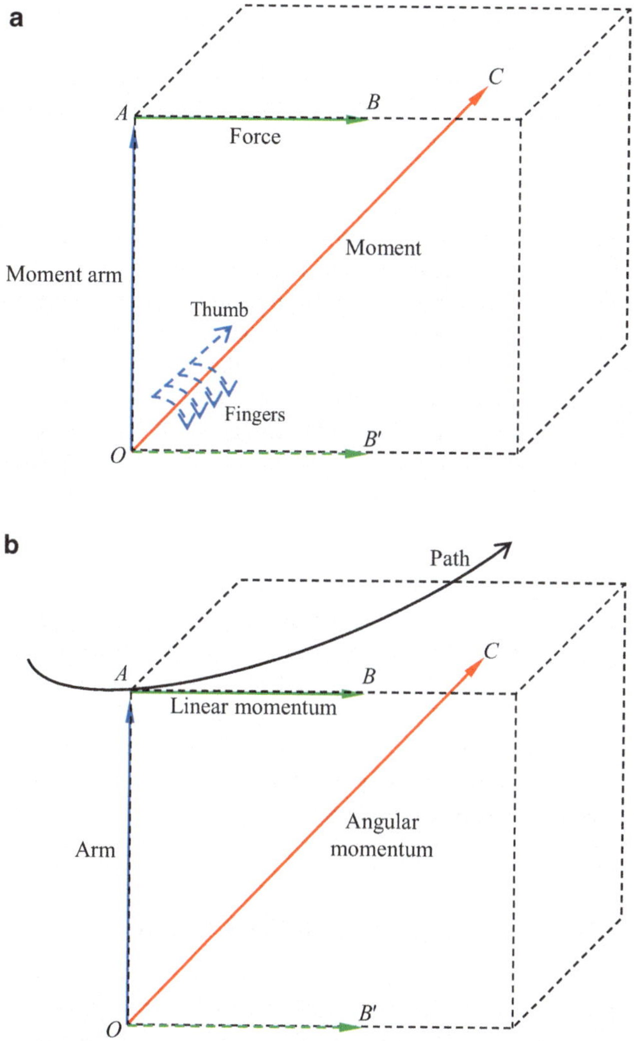

Fig. 1.2 Rotation about a point *O*. (**a**) Moment of a force, *AB*, about a point, *O*, by the right-hand rule. (**b**) Angular momentum about a point, *O*, of a particle taken at the point *A*, where its velocity is along *AB*. Before applying the right-hand rule to determine the direction of the final vector, *OC*, the vectors, *OA* and *OB*′, must have a common origin at *O*

to pass through the point *O*. For example, as shown in Fig. 1.2a, if a horizontal force is applied along *AB* at the end, *A*, of a vertical stick, *OA*, then a moment is produced about its other end, *O*, such that the moment arm is equal to the length of the stick, and the moment acts toward another horizontal direction, OC. The right-hand rule is applied by transferring the force vector, AB, to a parallel vector, OB', which passes through O, such that all the three vectors have the same origin, O.

A particle traveling with a velocity has an *angular momentum* about a given point O, whose magnitude is defined as the product of the particle's mass, its speed, and the arm length measured perpendicular to the direction of motion from the point O. Since the product of the mass and the velocity of a particle is termed its *linear momentum*, the magnitude of the angular momentum is thus the product of the magnitude of the linear momentum and the arm length. Figure 1.2b depicts the angular momentum vector of a particle traveling along a path at a point A where its velocity is along AB. The arm is measured in the direction OA, while the angular momentum vector has the direction OC given by the right-hand rule.

An object occupying the space is called a *body*, and consists of a large number of particles, each of which can have different positions, velocities, and accelerations in a particular reference frame. If the distance between every two particles on a body remains constant with time, such a body is said to be *rigid*.

The mass of an solid object divided by its volume is called its *density*. If a solid object is in contact with a fluid (either liquid or gas), it experiences a force on its surface due to the interaction of the fluid particles. This force, when divided by the area of the contacting surface, is called *pressure* exerted by the fluid. If the contacting surface area shrinks to a point (infinitesimal area), the pressure of a fluid at that point is obtained. Similarly, we can talk about the density of a fluid at a given point by considering its mass in the limit that the volume is decreased to a very small value (almost zero) at that point. The pressure and density of a fluid, as well as its temperature, can vary from point to point. We can feel this variation, for example, when a heater is turned on in a closed room.

1.3 Archimedes Principle and Balloon Flight

The earliest form of sustained human flight was in lighter-than-air vehicles, which derive their support in the air (called *lift*) from the force of buoyancy. This is the principle whose discovery had Archimedes running out of his bath crying "Eureka, eureka," and essentially requires a vehicle to displace a larger mass of air than its own mass.

Archimedes principle states that the buoyant force experienced by an object is equal to the weight of the surrounding fluid displaced by it. The same force that enables a heavy ship to float on water due to its large displacement is responsible for the flight of balloons, airships, and dirigibles. If a balloon is filled with a light gas such as helium—or if the air inside it is heated up—it expands to occupy a larger volume than it otherwise would if it were only filled with the air surrounding it. In so doing, it displaces a larger weight of air than its own weight, and hence rises up. It then continues to rise until it reaches an altitude (or height) where the density inside and outside the balloon becomes the same.

We note that the atmosphere is a large body of air bound to the earth by gravity, and is also subject to Archimedes principle. The less dense layers of the atmosphere ride atop the heavier layers, until an equilibrium of the buoyant force and the force of gravity prevails. That is why we find a smaller atmospheric density (and pressure) as we climb a mountain. The ever decreasing density of air with the altitude brings the ascent of a balloon to stop at some point.

Archimedes principle had to wait for almost 2000 years before it could be applied to flight, which took the form of a hot-air balloon ascending some 200 ft above a Paris square in the late eighteenth century. However, this mode of flight is not very convenient for a fast and accurate transport, because the balloon can be driven to uncertain locations and altitudes by the ever changing atmospheric properties and winds. If one does not wish to be entirely at the mercy of the atmospheric conditions, a different mode of flight must be sought that does not entirely rely on buoyancy.

The *aerostatic* (or LTA) category of flight vehicles includes the hot-air balloons, blimps, and airships. While the balloons have no way of propelling them forward, the blimps and airships have engines driving the airscrews (or *propellers*) as propulsion devices. Furthermore, as an airship moves forward, it can be stabilized and controlled by the help of the air flowing past a wing-like tail surface. In order to understand how the propulsion, stability, and control mechanisms work, a separate science called *aerodynamics* is required, which is briefly introduced in the next section.

1.4 Aerodynamic Flight

Since aerostatic lift requires a large volume of a light gas (such as hot air, hydrogen, or helium) to be either created or carried along with the LTA vehicle, the resulting design of a balloon or an airship tends to be quite bulky. The flight speeds of an aerostatic vehicle are quite small due to the huge resistance provided by the vehicle (called the *drag* force) as it moves through the atmosphere. Furthermore, the propulsion and control of an airship are poor, especially in the presence of strong winds.

In order to improve the speed of flight, as well as to make the flight more controllable a bird-like flight mechanism has always been sought. As birds can take-off, land, soar, and fly over long distances, seemingly at will, without requiring a bulky flight apparatus, they provided the ultimate inspiration for the design of a similar vehicle for manned flight. In order to achieve this feat, it is first necessary to understand how a heavier-than-air object like a bird can generate lift from the atmosphere. It is clear that the birds can fly by flapping their wings in the air. Hence the science of studying the mechanism by which lift is created by moving an object relative to the air came to be known as *aerodynamics* (as opposed to aerostatics discussed earlier). While a manned vehicle which fully makes use of a flapping mechanism (called an *ornithopter*) has yet to be designed, much simpler aerodynamic lifting mechanisms have been successfully implemented over the last century. These are the fixed-wing flight of gliders and airplanes first devised by the Wright brothers, and the rotary-wing flight of gyrocopters and helicopters.

The construction of aerodynamic flight vehicles consists of either a fixed, or a rotating wing (called rotor) mounted symmetrically on a body (called the *fuselage*). The propulsion in an airplane is provided by an engine which creates a forward propelling force called the *thrust* via a separate aerodynamic mechanism (comprising either propellers or jet exhaust) powered by a suitable engine. The propulsion device is usually fixed relative to the airplane, hence the thrust is provided along a fixed direction. In a helicopter, the propulsion mechanism is not separate, but involves tilting the rotor to produce a thrust in the desired direction. Thus, while an airplane can only move forward, a helicopter can fly sideways, backwards, or even remain stationary in the air (hover). The lift and thrust allow airplanes and helicopters to maintain constant speed and altitude of flight. A glider and a gyrocopter do not have any engines, and thus cannot maintain a constant speed and altitude while flying. This is because a flight vehicle always experiences an aerodynamic drag while moving through the atmosphere. Unless energy is supplied by the engines to overcome the drag, the speed, and altitude of a vehicle cannot be maintained for any length of time.

1.5 Basic Laws of Motion

Before embarking on the study of flight mechanisms, let us consider the basic laws of motion which are applied to any moving, solid object. In this section, the moving object is idealized to be a point, hence it has no physical dimensions. The motion of a point in a three-dimensional space is termed *translation*. Isaac Newton proposed the following three basic laws of motion in his revolutionary work dated 1687, titled *PhilosophiæNaturalis Principia Mathematica* ("Mathematical Principles of Natural Philosophy"):

1. A particle moving in a *straight line* at a constant *speed* will continue to do so unless a *force* acts upon it.
2. The time *rate* of change of the *velocity* (i.e., the *acceleration*) of a particle is directly proportional to the force acting upon it. The constant of proportionality is the *mass* of the object. By defining the product of the mass of the particle with its velocity as its *linear momentum*, it is implied that the time rate of change of the linear momentum vector is equal to the force applied to the particle.
3. If a particle exerts a force on another particle, the second particle simultaneously exerts a force equal in magnitude and opposite in direction on the first. The second force is called a reaction to the first force.

Newton's laws are valid only for an *inertial* observer. Such an observer, by definition, must not be undergoing any acceleration; otherwise the velocity and acceleration measured by her would not satisfy Newton's laws.

Newton's first law of motion implies that the velocity of a moving particle is a vector whose magnitude (speed) and direction (straight-line motion) remain constant with time, until and unless an external agency (force) acts upon it. This is the basic property (inertia) possessed by all objects having mass. Many common flight situations involve a straight-line motion of the vehicle at a nearly constant speed (e.g., an airplane flying steadily and level or climbing at a constant rate, a glider descending at a fixed glide angle, etc.). In such a flight, all the force vectors on the vehicle (regarded as a particle) must add to zero, otherwise, by the first law, either the speed or the direction of the vehicle will change with time. As we shall see later, the equilibrium condition where all the forces on a vehicle are balanced (i.e., sum to zero) is a natural atmospheric flight state, requiring the smallest energy expenditure for a given range. However, such an equilibrium state does not exist for space flight vehicles, which are always pulled toward the center of a massive body by gravity. Similarly, an airplane either taking-off, or performing maneuvers, has an accelerated flight condition where the forces are unbalanced.

The second law of motion tells us how a change in the velocity vector is produced when a force is applied. Suppose the force is applied only tangentially (that is, along the direction of motion). Then there cannot be any change in the direction of the motion, because the initial velocity and its change due to the applied force are both along the same straight line. This is clear from the vector diagram of Fig. 1.1b in which all the three vectors (the initial vector, AB, the change in the vector, BC, and the final vector, AC) are all lined up, hence the triangle collapses into a straight line. The magnitude of the final vector in such a case can be obtained by a simple arithmetic addition or subtraction, depending upon whether the force is applied along, or opposite to the fixed direction of motion. For example, a tangential force acts on a car when its brakes are applied, causing its speed (magnitude of velocity vector) to decrease with time. The acceleration in this case is a scalar quantity, which is simply obtained by measuring the rate of change of the speed with time. Furthermore, the third law dictates that when the brakes are applied to a car, it in turn applies an equal and opposite force on the occupants, causing them to be pushed forward.

Now, let us consider a case where the magnitude of the velocity is unchanged, but its direction is varied. This can happen if the force is applied to the object at a 90° angle (or normal) to the direction of initial motion. In such a case, there is no possibility of a change in the magnitude of the velocity vector (i.e., its speed), but only in its direction. This is the case where both the initial and the final velocity vectors have an equal length, but different directions, and the vector triangle becomes an isosceles triangle (as shown in Fig. 1.1c). A flight example where the applied force is always normal to the direction of motion is that of a satellite in a circular orbit around a spherical planet. Here, the gravity is the only external force, and pulls the satellite toward the center of the planet, leaving the speed of the satellite constant with time.

Whenever the flight direction is changed without any change in the speed, a force must have been applied to the vehicle along the instantaneous radius of curvature of the flight path. Since such a force—called the *centripetal force*—acts in a direction perpendicular to the instantaneous flight direction, it is incapable of changing the speed.

The second law of motion encompasses the most general situation, namely when the force is applied to the object at an arbitrary angle to the direction of motion. The resulting change (i.e., acceleration) is then both in the magnitude (speed) and the direction of the velocity vector, and the object describes a curve with a varying speed. An example of such a motion is a roller-coaster ride which many readers may have enjoyed, and constitutes variations in both the speed and the direction. A general flight path through the air or space is similarly analyzed by Newton's second law of motion.

The third law of motion allows us to easily understand the source of many forces, which would otherwise require a detailed mathematical analysis. A complex interplay of various mechanisms often boils down to a simple action–reaction pair of forces, to which the third law can be applied. For example, later in this book we shall apply the third law to explain the creation of the aerodynamic lift in fixed, rotary, and flapping-wing flight. In formal textbooks on aerodynamics, the lift is explained by more complicated fluid mechanics principles. Similarly, the creation of the thrust by an aircraft or a rocket engine is directly explained by Newton's third law.

1.6 Translation in a Curve

Now we are in a position to understand very broadly the difference between straight-line (rectilinear) motion and curved (curvilinear) motion. To summarize the earlier discussion, a force must be applied in order to cause a change in the velocity of a particle. If the direction of the applied force is along (or opposite) to the direction of motion, it can only cause a change in the speed of the particle, but not in its direction, hence the object continues to move in a straight line. Whenever a force is applied at an angle to the direction of initial motion, a change in the direction is always produced, resulting in a curved path. The essential nature of a general curvilinear motion is depicted in Fig. 1.3. At a given instantaneous position, A, along the trajectory of a particle, there exists a centripetal acceleration, AB, which acts toward the instantaneous center, O, of an imaginary circle. The imaginary circle is constructed such that a tangent to it at the point A gives the instantaneous velocity vector, AC. The angle $\angle OAC$ is 90°, implying that the centripetal acceleration acts normal to the flight path. The radius of curvature of the path at a given instant is the radius of the imaginary circle, which may keep on varying with time if there also exists a net force in the tangential direction (i.e., along the trajectory), causing a change in the speed of motion.

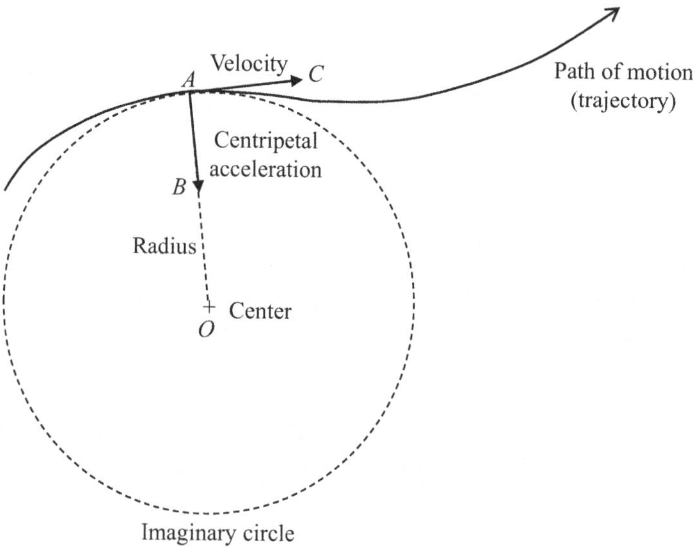

Fig. 1.3 Curvilinear motion in a plane. A particle follows a curved path if at every point, there exists a force component normal to the direction of motion, called the centripetal force. At a given point *A* along the trajectory, the centripetal acceleration, *AB*, acts toward the center *O* of an imaginary circle, a tangent to which at *A* gives the velocity vector, *AC*. The radius of curvature of the path at a given instant is the radius of the imaginary circle, which may keep on varying with time

Let us consider the simplest case of curved motion, namely a particle moving in a circular path. In this special case, the force is applied continuously toward the center (i.e., in the centripetal direction) of a fixed circle traced by the point mass. If we break down the time into many infinitesimal intervals (or instants), then at each instant, it is as if the force is applied at a 90° angle (i.e., normal) to the instantaneous direction of motion. Consequently, there is no change in the speed of the object. However, the direction keeps on changing at every instant due to the normally applied force, and the result is a circular motion. The centripetal force must always act in order to maintain such a motion.

By the second law of motion, the magnitude of the force increases in proportion with the acceleration, that is, the time rate of change of velocity. In a circular motion, the acceleration is proportional to the square of the angular speed (revolutions per second). Now, let us also consider how the centripetal force is applied to the particle. Suppose there is another object (or person) applying the centripetal force on the moving particle. When children play with a rock tied to a string, they make it move in a circle. At every instant, the person holding the string applies a pulling force on the moving rock. However, by the third law of motion, the rock applies an equal and *opposite* force (i.e., the *centrifugal force*) on the person, and we feel the tug on the string which keeps it straight and taut. Now suppose in order to make the rock move faster, the person applies such a large force that the string breaks at a point in the circular motion. The rock will immediately depart from a circular motion at this point. If there is no other force acting on the rock (such as the force of gravity, or an atmospheric force), then it will then continue to move in the straight line which is tangential to the original circle. This is the well-known *slingshot effect* by which objects can be hurled as projectiles by first rotating them in a circle. Of course, once the projectile is launched, it moves under the influence of gravity and atmospheric forces in a curve, rather than a straight line.

By the first law of motion, it is clear that the only time a particle experiences a force-free condition is when it is moving in a straight line at a constant speed (uniform rectilinear motion). In contrast, at every instant that a particle moves in a curve, its speed may (or may not) be constant, but the direction is instantaneously changing due to a normally applied centripetal force. However, it resists the curved motion by applying a normal force in the opposite direction. Whenever such a force cannot be supplied, the curved motion is no longer possible, and the object flies away in a straight line. This is the principle behind the breaking strings, and disintegrating discs when required to move at a high rotational rate, but without supplying the necessary centripetal force by a sufficiently strong structure.

Here we are touching on an important but usually difficult to grasp principle. The way an object can be forced to move depends only upon the forces that can be applied to the object. When we try to move a ball around a desired curve formed by a groove, we always try to supply the necessary initial velocity. If the initial speed is too small, the ball does not get into the required groove, and exits. However, if the initial speed is too large, the necessary normal force cannot be supplied by the groove, and the ball escapes the groove. The speed of the object, as well as the rate of turn it is required to make determines the required acceleration, which must obey the second law of motion in terms of the available normal force. For a given flight speed, the smaller the radius of the required turn, the larger centripetal acceleration (hence normal force) is required. For a given radius of turn in a curve, a larger speed requires a proportionally larger centripetal acceleration, and thus a larger normal force. In flight, the normal force is usually supplied by the lift of the wings. If an aircraft is moved in a much tighter curve, or at a much higher speed than for which it is designed, then the lift produced by the wings can exceed their structural strength, causing them to disintegrate. On the other hand, a human pilot can "black-out" in a tight maneuver, where the normal acceleration is so high that the blood is prevented from circulating to the brain due to the limitation of the heart's pumping capacity. Therefore, the acceleration limits of an airplane are usually placarded on the instrument panel in the cockpit, so that they are never exceeded for safety reasons.

1.7 Coriolis Acceleration

When a particle flies along a curved path with a varying speed, it experiences an acceleration which is the vector sum of the acceleration normal to the flight path, and the tangential acceleration along the flight direction. However, since the flight direction is changing continuously, so are the "normal" and the "tangential" directions, and a complicated picture of the acceleration being experienced by the particle is thus presented to an observer who is not herself accelerating. Because Newton's laws are stated with respect to an inertial (i.e., non-accelerating) observer, it is only such an observer who can claim to understand the motion of the particle. Let us see what an inertial observer might say about a general, accelerated flight. For simplicity, we assume that the observer is at rest, and the motion takes place in a fixed plane.

Consider two mutually perpendicular axes of constant lengths, OA and OB, attached to a particle denoted by the point O as it moves in a plane. The axis OA is chosen such that it is always parallel to the velocity vector, whereas the axis OB is normal to the velocity vector. After a very brief time interval, the velocity has changed both in magnitude and direction, from an initial value, OC to a final value, OC'. Hence, the new orientations of the two reference axes are now OA' and OB', respectively. But their respective lengths are unchanged because they are taken to be constant with time.

The stationary observer sees the velocity vector has changed from OC to OC', therefore it must have undergone an acceleration, whose direction is along the net velocity change, CC', as shown in Fig. 1.4. The net velocity change is then expressed as the vector sum of the tangential component, CD, and the normal component, CE, which correspond to the change in the length of the original velocity vector, OC, and the change in its direction, respectively, such that the new velocity vector, OC' is generated. If there was no change in the speed involved, the net acceleration would be only a centripetal acceleration along the vector CF, such that the lengths of the velocity vectors OC and OF are equal (Fig. 1.4a). In that case, the new velocity vector, OC' would be the same as OF.

But the centripetal change, CF, and the tangential velocity change, CD, *do not* add up to the net velocity change, CC', as shown in Fig. 1.4b. This is because the net normal acceleration, proportional to CE, is different from the centripetal acceleration, which is proportional to CF. Hence, an additional vector, FE, must be added to the centripetal velocity change, in order to produce the required net change, CE, as shown in Fig. 1.4c. The difference between the net normal acceleration and the centripetal acceleration is called the *Coriolis acceleration*, which is produced whenever the acceleration is resolved in a rotating reference frame (such as the frame OAB). The magnitude of the Coriolis acceleration is proportional to the product of the speed and the rate of rotation. In the present case, the frame is fixed to the velocity vector, and thus rotates with it at the same rate.

Note that if an observer were to be moving with the *same* velocity and acceleration as the point O, she would see the net acceleration very differently from the stationary observer. From the new observer's perspective, the frames OAB and $OA'B'$ would be identical and unchanged with time. Consequently, the net velocity change is then resolved by the observer in the moving reference frame, $OA'B'$, comprising a tangential component CD' parallel to OA', and the normal component, CE', parallel to OB'. The only acceleration directly noticed by the observer would be that corresponding to the tangential velocity component, CD', because such an

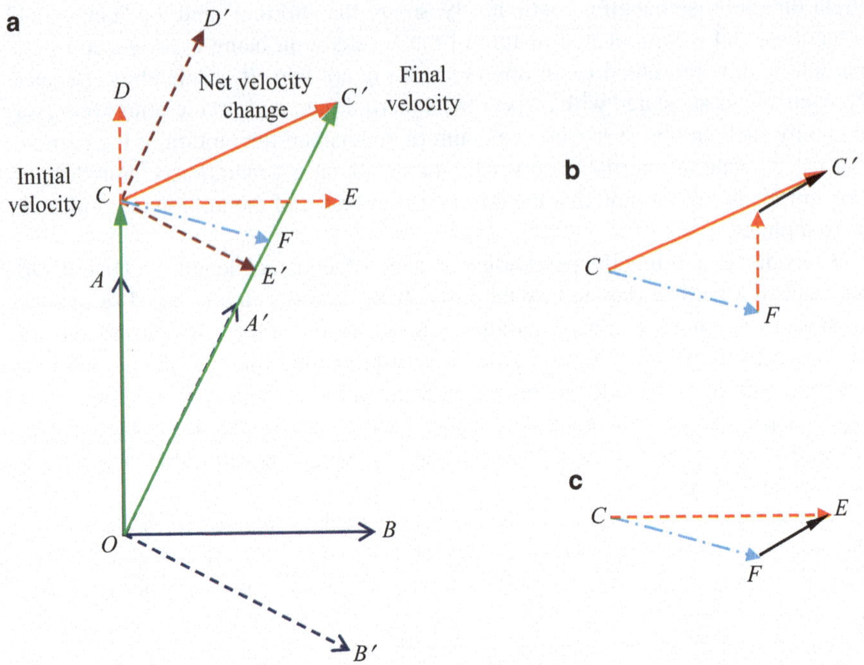

Fig. 1.4 (**a**) The source of Coriolis acceleration in an accelerated flight resolved in a rotating reference frame. In an infinitesimal duration, there is a change in both the magnitude and the direction of the velocity vector, *OC* to *OC*′, which is resolved in the rotating frame, *OAB*, by a stationary observer. (**b**) The centripetal velocity change, *CF*, and the tangential component, *CD*, do not add up to the net velocity change, *CC*′, due to the missing Coriolis vector. (**c**) The difference between the net normal acceleration (along *CE*), and the centripetal acceleration, (along *CF*), is the Coriolis acceleration, (along *FE*), produced by the rotation of the reference frame

observer has no knowledge of the rotation of her own reference frame. If such an observer were in a flight vehicle, she would also experience a force exerted upon her by the vehicle in a direction opposite to the vector *CE*′. Such a force would lead her to believe that the vehicle is undergoing a *centrifugal* (away from the center) acceleration. But this would give us a false picture, because the observation is not carried out in an inertial reference frame.

Coriolis acceleration is experienced by a person moving radially on a turn-table, which is rotating at a constant rate. Since such a person has different inertial speeds and directions at different times, both centripetal and Coriolis accelerations are applied to her. All objects traveling on the earth's surface similarly experience a Coriolis acceleration due to the earth's rotation. Aircraft flight resolved in a ground-based reference frame is similarly seen to have a Coriolis acceleration. Since the earth's rotational rate is quite small, and air travel takes place at low speeds, the Coriolis acceleration thus produced is negligible. However, if the flight duration is of several hours, Coriolis acceleration can produce a significant error in estimation of the flight path. Spacecraft and rockets—due to their much higher velocity—have a significant Coriolis acceleration caused by the earth's rotation, when their flight is analyzed in a ground-based reference frame. If the same motion is referred to in an inertial (non-rotating) reference frame, there would not be any Coriolis acceleration.

In our explanation of the laws of motion by vector diagrams, we have considered only an *instantaneous* change in the velocity vector. This is the special case when the applied force acts only instantaneously (that is, for a very tiny duration) and then stops acting. Such a force is called an *impulsive* force, an example of which is the force applied to a nail by a hammer, or to a ball by a baseball bat. Since the time interval during which the force acts on the object is very small (or infinitesimal), it is reasonable to assume that the change in the motion takes place in a zero time (or instantaneously). However, when forces act over a finite time interval, one has to carefully *integrate* (or sum) all the changes that have happened over infinitesimal durations. This requires a special branch of mathematics called *calculus* (also invented by Newton) to accurately determine the changes happening over finite durations. It is not our purpose to carry out such calculations in this book.

1.8 Rigid-Body Dynamics

Newton's laws of motion can be applied to a body by considering a vector sum of the individual motion of all the particles of the body. The application of Newton's second law to a body is a summation of the accelerations and the forces acting on every particle. If there are no electromagnetic fields present inside a body which can defy Newton's third law of motion, the internal forces between every two particles of the body form an equal and opposite, action–reaction pair and their vector sum is zero. Hence, only the external forces can cause a change in the motion of a body. This summation yields the following interesting properties of motion:

1. Each body has a special point associated with it called the *center of mass*. The *translational* motion of the body is defined to be the motion of its center of mass.
2. The center of mass moves exactly as if all the mass of the body is concentrated on it. All the external forces applied at the various points of the body are summed to a net external force applied to the center of mass. This net force equals the mass of the body, multiplied with the acceleration of the center of mass (also the time rate of change of the linear momentum of the center of mass).

The translational motion of the center of mass gives an important information about its motion. If the body were to be approximated by a particle, then its motion would be given by the motion of the center of mass. When we analyze the motion of a flight vehicle driven by the various external forces, we first consider its translational motion only. This aspect of the aircraft and rocket flight is termed the vehicle's *performance*.

However, the translational motion is not a complete description of how a body can move. Since the external forces act at the various points of a body, they produce a rotation of the body that must also be explained. For example, a car's motion can be only partially described by how its center of mass travels in three-dimensional space. It is also necessary to describe how the vehicle rotates about a vertical axis while being steered on the road. More accurate information of a car's motion can be obtained if we consider the motion of its individual parts, such as the rotation of the tires, the motion of the crankshaft, the opening and closing of the valves, etc. Still more accurate description of the car's dynamics can be obtained by considering the structural deformation of the tires, the vibration of the suspension and chassis, and the noise produced by its engine. The detailed description of the various parts of the car might not really be necessary if only its motion on the road has to be described, but will be essential if the stresses produced on its structure, and the effects of its motion on the passengers are to be analyzed. Thus, we focus only on those aspects of a vehicle's motion which are necessary for the limited objectives. Such an approximation which requires a neglect of some aspects of the motion under study, while focusing on the principal objectives, is called *idealization*. An example of idealization is the approximation of a real feature by a simple geometrical shape, such as treating the earth to be a perfect sphere, the tires to be exactly circular, a flight path to be a straight line, etc. Another example of idealization is treating an object as if its can move (i.e., change its position and velocity) only in a limited number of ways, called the *degrees of freedom*. Hence, while a car actually has infinitely many degrees of freedom, we only need to consider its forward and sideways translation, as well the rotation about a vertical axis which is fixed to the car and moves with it. This gives us a three degree-of-freedom model for the car's motion, and neglects the relative motion of its various parts. Such an idealization essentially treats the car to be a rigid body, which greatly simplifies its model.

The rigid-body assumption is based on neglecting both the structural deformation of a vehicle's structure and the relative rotation of its parts. When we apply Newton's second law of motion to each particle of a rigid body, and take a summation of the moments about an arbitrary point, O, we arrive at the following additional features of its motion:

3. The summation of the moments of all the external forces acting on a rigid body about an arbitrary point O result in a net *external torque* acting about O.
4. If the point O is either a fixed point, or the center of mass of the rigid body, then the net external torque equals the time rate of change of its net *angular momentum* about the point O. The angular momentum of the rigid body is defined as the vector sum of the angular momenta of all its particles about the given point. The change of angular momentum of a rigid body under the action of a net external torque produces a variation of its *angular velocity* about an axis passing through the point O, and in the direction of the applied external torque. If there is no external torque applied to a rigid body, then its angular momentum about the center of mass is conserved.

The point O is conveniently taken for a flight vehicle to be its center of mass. Note that both the torque and the angular momentum are vectors, i.e., they have magnitude as well as direction. They are produced by taking a vector sum of the individual external moments and angular momenta, respectively, of all the particles. The direction of the torque gives the axis about which the rotation of the rigid body is produced.

The translational and rotational dynamics of a rigid body are pictorially depicted in Fig. 1.5. The translational dynamics (Fig. 1.5a) is described by the position and velocity of its center of mass, *O*, relative to an inertial reference frame, *SXYZ*. A summation of all elemental forces of the body produces a net external force applied at *O*. The rotational dynamics (Fig. 1.5b) about the center of mass, *O*, is measured by the changing orientation of a coordinate frame, *Oxyz*, rigidly fixed to the body with origin at *O*. This measurement is carried out relative to the inertial reference frame brought to coincide its origin with the point *O*. The time rate of change of the angular velocity of the frame *Oxyz* relative to the frame *OXYZ* is determined by the net external torque acting about *O*. The angular velocity thus plays the same role in the rotation of a rigid body as the velocity of translational motion.

The difference between translational and rotational motions can be understood by considering the motion of a rigid body whose center of mass moves in a straight line at a constant speed. Since all the particles of the body are moving with the same speed in parallel straight lines, there is neither an angular velocity, nor angular momentum of the rigid body about its center of mass (or about any fixed point). Therefore, there is no torque acting on the body, and it is said to be in a pure translation. In contrast, consider a rigid body whose center of mass is at rest, but all the particles are describing circles about the center of mass. Such a body is said to be in a pure rotational motion. Any motion in which the center of mass moves in a curve is both translating and rotating.

The use of a body-fixed frame, *Oxyz*, makes the rotational inertia of the rigid body to be constant with time. However, the rotational inertia is no longer a scalar—such as the mass in the translational motion—but becomes a mathematical quantity called a *tensor* (or a square matrix with three rows and three columns) for the general rotational motion shown in Fig. 1.5b. A matrix product of the inertia tensor with the angular velocity vector gives the angular momentum vector. Hence, the inertia tensor can be thought of as a transformation which changes both the magnitude and the direction of the angular velocity vector to produce the angular momentum.

The simplest rotation is the one where the applied torque and the angular velocity vectors are aligned (i.e., they act about the same axis passing through the center of mass). This is termed a *single-axis rotation*. If two (or more) single-axis rotations are involved, they can be added vectorially to produce the rate of change of the angular momentum. In such a case, the angular velocity is multiplied by a scalar called the *moment of inertia* about the given axis of rotation to yield the angular momentum. Since both angular momentum and angular velocity have a common direction, they are related by a simple multiplication. Newton's second law of motion applied to a single-axis rotation states that the time rate of change of the angular velocity is proportional to the applied torque, and the constant of proportionality is the moment of inertia of the body about the concerned axis.

The single-axis rotation at a constant rate gives a constant angular momentum, and is therefore an equilibrium state where no external torque is applied. This

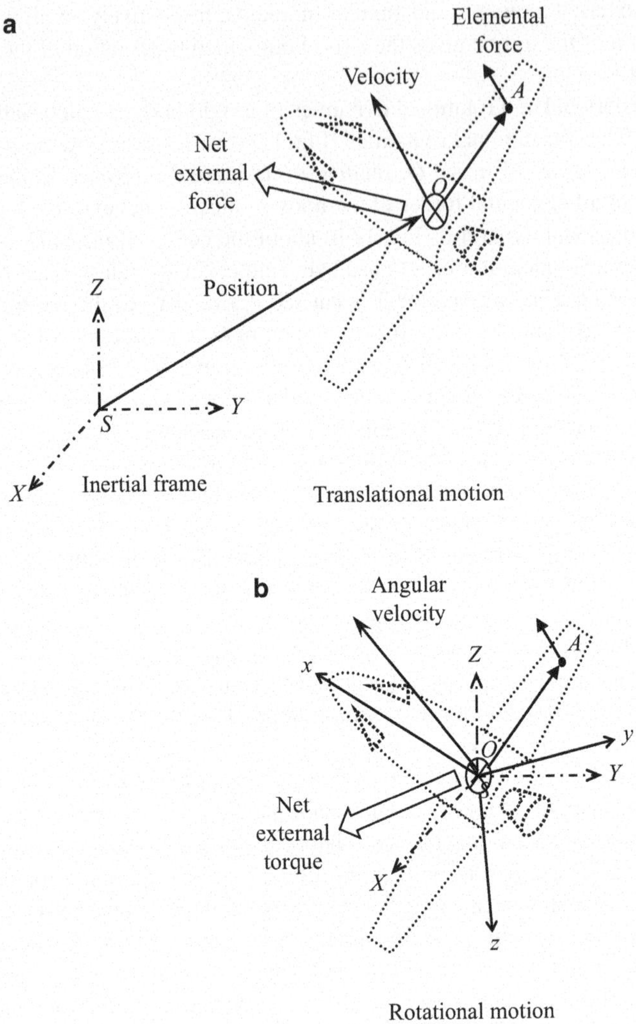

Fig. 1.5 (**a**) The translational dynamics of a rigid body is described by the motion of its center of mass, O, relative to an inertial reference frame, $SXYZ$. A summation of all elemental forces of the body lead to a net external force applied at O, whose velocity and position change according to the acceleration thus produced. (**b**) The rotation of a rigid body about its center of mass, O, is measured by the orientation of a coordinate frame, $Oxyz$, fixed to the body with origin at O, relative to the inertial reference frame brought to coincide its origin with O. The time rate of change of the angular velocity of $Oxyz$ relative to $OXYZ$ is determined by the net external torque acting about O

equilibrium state of rotation is analogous to the straight-line translation at a constant speed, where the linear momentum is constant, and no external force acts on the body.

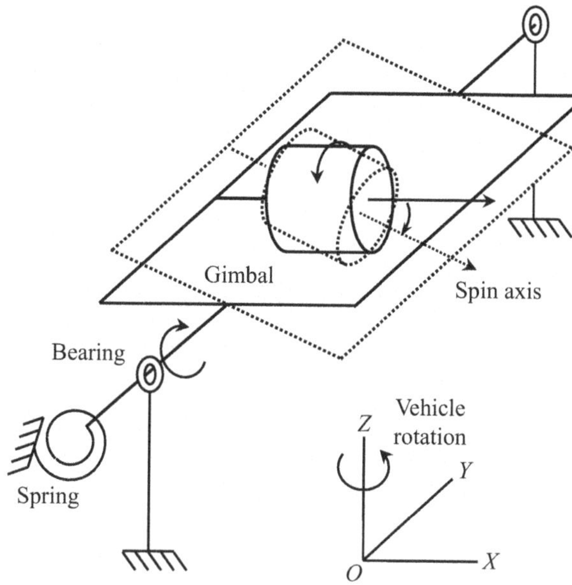

Fig. 1.6 A rate gyro works on the principle of conservation of angular momentum. A rotor spinning about an axis, *OX*, at a constant speed is made to pivot about a perpendicular gimbal axis, *OY*, whenever a rate of rotation about the third axis, *OZ*, is sensed. The restraining spring makes the gimbal angle proportional to the sensed rate of rotation, whereas the gimbal bearing absorbs any transverse torques

A common application of the principle of angular momentum conservation is in gyroscopic devices, which are often employed for controlling the rotational motion of flight vehicles. A *gyroscope* (abbreviated as *gyro*) consists of a rotor spinning about an axis passing through its center of mass, and mounted such that the spin axis can be tilted about another axis (called the *gimbal*) which is perpendicular to it. The rotation of the spin axis about the gimbal is restrained by a spring (Fig. 1.6), and can also be connected to a variable electrical resistance (called a *potentiometer*), such that any rotation about the gimbal produces an electrical signal (current or voltage). Any change in the spin axis can thus be measured and calibrated. If the vehicle on which the gyro is mounted undergoes a rotation about an axis normal to both the spin axis and the gimbal, then a resisting torque is generated by the rotor about the gimbal axis in order to oppose the change in the angular momentum. There are also small transverse torques created about two transverse axes normal to the gimbal. However, while the torque about the gimbal causes a gimbal rotation, the transverse torques are absorbed by the gimbal bearing. The presence of the restraining spring produces a gimbal angle proportional to the vehicle's rotation rate, which can be measured. Hence, such a sensor can be used to sense the rotational rate of the vehicle normal to both spin axis and the gimbal, and is called a *rate gyro*. If the restraining spring is removed, then the gimbal displacement thus produced is directly proportional to the net angular rotation of the vehicle about the given axis. This new device is called a *rate-integrating gyro* because it gives the time-integral of the rate, i.e., the net angular rotation. As will be discussed in the later chapters, both rate gyro and rate-integrating gyro are useful sensors of a vehicle's single-axis rotation.

A detailed discussion of the general (multi-axis) rotational dynamics is beyond our scope, because it requires a mathematical description of the inertia tensor. Since a sequence of single-axis rotations about different perpendicular axes can be used to produce a general rotation, all explanations of the rotational dynamics in this book will employ such a sequence.

1.9 Flight Stability and Control

As depicted in Fig. 1.5, a flight vehicle is free to move forward, sideways, up and down, as well as to rotate about three mutually perpendicular axes of a reference frame, *Oxyz*, fixed to its body. These different ways in which a vehicle can move are called its *degrees of freedom*. A rigid vehicle thus has six degrees of freedom: the *forward*, *lateral*, and *vertical* translations of its center of mass along the axes *Ox*, *Oy*, and *Oz*, respectively, as well as the *roll*, *pitch*, and *yaw* rotations of the vehicle about the center of mass about *Ox*, *Oy*, and *Oz*, respectively (Fig. 1.5). If all the degrees of freedom of a vehicle are excited simultaneously and randomly, then its flight will be a chaotic motion. Neither the speed, nor the flight direction of such a craft would be manageable. Atmospheric flight is particularly susceptible to the vagaries of the aerodynamic forces, if the vehicle's attitude in the air is improperly managed. The reader might have experienced this when playing with a Frisbee or a paper dart in the childhood, and can appreciate how difficult it is to throw such an object such that it flies in a desired manner. Therefore, having a sufficient engine power and a lifting mechanism are not the sole ingredients for a successful flight.

It is important for any flight vehicle to have two important properties for making an orderly progress through the air or space: (a) stability and (b) controllability. *Stability* refers to the tendency of a vehicle to return to an equilibrium condition, once displaced from it by a disturbance. *Controllability* is the ability to change from one equilibrium flight condition to another as desired. Stability and control are achieved very differently in the aircraft, rockets, and spacecraft due to their different constructions, different modes of flight, and the different environments of operation.

Fixed-wing aircraft have a separate structure called the *empennage* for achieving stability and control. This consists of a set of small, wing-like surfaces mounted usually at the rear of the aircraft. A rear-mounted surface nearly parallel to the ground in a level flight is called the *horizontal tail*, while a similar surface mounted near the front of the aircraft is called the *canard*. The tail (or canard) stabilizes the upward/downward (pitching) rotation of the entire vehicle. Almost all designs have a large vertical tail to provide stability in the side-to-side rotation (yawing), which is necessary for maintaining a desired flight direction. Flight control is enabled by control surfaces. These are movable parts of the wing, the horizontal tail (or canard), and the vertical tail, and are called the *ailerons*, *elevator*, and the *rudder*, respectively. We will discuss the stability and control of aircraft in Chaps. 3 and 4.

Rotorcraft achieve stability and control through a tail rotor for yawing, and via the main rotor(s) for pitching and yawing (to be discussed briefly in Chap. 5). Spacecraft achieve their stability and control by rocket thrusters, rotors, gravity-gradient, and geomagnetic torques. Stability and control of spacecraft and rockets is covered in Chap. 6. The stability and control of rockets, discussed in Chap. 7, is based on rotating (gimbaling) of the engine nozzles for pitch and yaw, and via the aerodynamic fins for roll.

Chapter 2
Aerodynamics

2.1 Relative Air Flow

Aerodynamics refers to the study of forces applied to a solid object such as an airplane wing by a gas (typically air) flowing around it. Such a flow can be created either by moving the solid object through the atmosphere, or by blowing the air past a stationary object in a wind-tunnel. Since it is only the motion of the air relative to the object which creates the aerodynamic forces, it is unimportant how the flow has been generated as long its velocity and acceleration relative to the object are the same. For this reason, aerodynamic behavior of an airplane in actual flight can be understood by studying a model of a similar shape placed in a wind-tunnel. The relationship of the actual airplane's dimensions with those of the model is called scaling, which also affects the flow conditions (density, velocity, temperature, etc.) required for simulating the actual aerodynamic properties in a wind-tunnel test. The fluid being an infinite medium can have its properties changing from point to point. Let us define the flow properties as those of a tiny volume of fluid called a *fluid element*. This element can move by translation and rotation, and its shape can also deform due to internal stresses. It is useful to define the properties of the flow far upstream of the object. These are called the *freestream properties*. The relevant freestream properties which dictate the magnitudes of the aerodynamic forces experienced by an airplane wing are the *relative airspeed*, the *atmospheric density*, and *air temperature*.

When the relative airspeed is quite low, the local variations in the density caused by the air flow can be neglected, and we have what is called an *incompressible flow*. This is very much like the flow of a liquid which cannot be compressed by applying any amount of pressure, and transmits any pressure changes (called pressure waves) almost instantaneously everywhere. This concept is called *Pascal's law* and is the principle behind the working of hydraulic machines such as an automobile jack, where a small force can be used to lift a much heavier object by quickly transmitting the applied pressure through an incompressible hydraulic fluid. Since pressure is force per unit area (see Chap. 1), the force transmitted to the loaded end is much

A. Tewari, *Basic Flight Mechanics*, DOI 10.1007/978-3-319-30022-1_2

larger due to its greater area. However, it is important to maintain the applied pressure. If the hydraulic fluid were a compressible gas, some pressure would be inevitably lost in compressing the medium, therefore a smaller pressure would be transmitted for lifting the load. In addition to being inefficient, such a jack would also be a sluggish device, because the spring-like compression of the fluid would significantly delay the transmission of the pressure wave from one end to the other. Similarly, much smaller pressure differences would be created on an airplane wing by a flowing air, if the air behaves as if it is a compressible medium. Furthermore, the pressure waves crossing the air as a compressible medium would be significantly delayed.

2.1.1 Compressibility

The speed at which small pressure changes can be transmitted across a medium is called the *speed of sound*. The pressure waves are transmitted by making the particles of the medium vibrate along the direction of motion. When the material particles are tightly bound as in a solid medium, the energy comprising the pressure wave is quickly transmitted, and hence the speed of sound is high. On the other hand, when the medium particles are separated by large distances as in a gas, a small pressure wave would travel much more slowly due to the time taken by the particles to cross the intervening distances. Therefore, a denser medium has a higher speed of sound. The magnitude of the speed of sound would also depend upon the size and structure of the particles, which have to do with its chemical structure. Hence the speed of sound is related to the physical and chemical properties of the medium. In a perfectly incompressible medium, there would be hardly any change in the speed of sound caused by the flow of the medium because its density and chemical composition everywhere are the same. However, if the medium were compressible, it would experience density variations due to the flowing medium, and hence different points in the medium would have different values of the speed of sound. Since greater density variations are produced by a higher flow speed, it is natural to expect that a faster air flow would have larger changes in its local density. If the flow speed were to be so large that in addition to significant density variations, large changes in the temperature are also present which can affect the chemical properties of the medium, then we have an extreme case of compressible flow. Such flows are experienced by objects entering the atmosphere from the space due to their very high speeds. However, the speeds in normal airplane flight are seldom so high as to cause chemical changes in the air, and the latter can be regarded as a *chemically perfect* gas.

From the foregoing discussion, it is clear that the aerodynamic compressibility effects depend upon the flow speed relative to the speed of sound. To classify the flow regimes, it is thus important to define a parameter called the *Mach number* as the ratio of the freestream flow speed and the freestream value of the speed of sound. There are the following three categories of the flow, entirely depending upon the Mach number:

(a) Incompressible flow, where the Mach number is negligible (approximated to be zero).
(b) Compressible flow of a chemically perfect gas, where the Mach number is significant, but not so large as to cause variations in the chemical composition of the gas. This category is further divided into the following:

- Subsonic flow, where the Mach number is less than 1.
- Supersonic flow, where the Mach number is greater than 1.
- Transonic flow, where the Mach number is very close to 1.

(c) Hypersonic flow, where the Mach number is very much greater than 1. For air, this regime is defined by Mach numbers larger than 5. The hypersonic regime is usually accompanied by high temperatures, which cause the gas to depart from its constant (perfect) chemical properties found at lower Mach numbers.

In this chapter we will only consider the subsonic flow for simplicity, whereas the essential features of the hypersonic flow will be discussed in Chap. 6 in the context of atmospheric entry vehicles.

2.2 Lift and Drag

Lift and drag are the basic aerodynamic forces created whenever a solid object moves in the atmosphere in a fixed direction. As we have seen previously, air consists of a large number of tiny particles called molecules. These molecules are always in a random motion (called Brownian motion) even though a volume of the air (such as in a room) is at rest. The particles being tiny cannot be observed, but their effects are felt by us two distinct ways. One way in which we can feel the air particles is by the atmospheric pressure exerted by them on the skin. This is the force per unit area acting normal to our body surface at a given point. If the object is at rest, all points on its surface have the same pressure, because the random motion of molecules causes them to strike the stationary object in the same way at all points. However, this changes as soon as motion takes place relative to the air. For example, when we move a hand through the air, we immediately feel the increased pressure on the hand in a direction opposite to that of the motion. This is the aerodynamic drag force which arises due to the motion of the object relative to the air, and *opposes* the motion. We can also feel aerodynamic drag when we stand in a breeze, or directly in front of a fan, in the form of a light brush of air against the part of the body surface which is tangential to the flow of air. This is the second source of drag caused by the brushing (friction) of the molecules against our skin, and is quite different from the impact of the molecules normal to the skin. Thus we have two sources of aerodynamic drag, namely, the pressure and the skin friction.

Lift is the name given to the net aerodynamic force acting *normal* to the direction of the relative motion to the air which takes place in a constant plane. It is the force which makes airplane flight possible, because it can be made to counter gravity by devising a proper lifting mechanism (wing). As in the case of the drag, the lift also has two fundamental sources, viz. the pressure and the skin friction.

When the motion of an object through the air is not only confined to a constant plane, but also involves a sideways (or lateral) motion, it also experiences a third force called the *sideforce* which acts in a lateral direction, normal to the flight path. The lift, the drag, and the sideforce are thus mutually perpendicular forces.

In addition to the aerodynamic forces of lift, drag, and sideforce, an object moving through the air also experiences an external torque, that can be resolved into moments about three mutually perpendicular axes, $Oxyz$, fixed to the body (see Fig. 1.5b). These are the *rolling moment* about Ox, the *pitching moment* about Oy, and the *yawing moment* about Oz.

From the foregoing discussion, it is clear that all the aerodynamic forces and moments are caused by the following two basic effects:

- Air particles striking the surface of the solid object, such that an exchange of momentum takes place in a direction normal to the surface. This rate of change of momentum per unit area normal to the surface is called the pressure.
- Air particles sticking to, or dragging along the solid surface, such that an exchange of momentum takes place in a direction tangential to the surface. This rate of change of momentum per unit area tangential to the surface is called the shear stress. The skin friction force experienced by a solid object is the integration (summation) of the shear stress acting at all the points on the surface.

Every object experiences both of these effects, and the magnitude of each effect depends upon the following factors:

- The shape presented by the object to the oncoming airflow.
- Properties of the atmosphere prevailing at that point.
- The flight speed of the object relative to the atmosphere.
- The size of the object indicated by a reference length and a reference area.

For any flight object such as an airplane, the shape is selected to maximize the lift to drag ratio in the normal cruising flight. Of course, there are other design requirements, such as the volume required to house the payload, the engines, the structure, the fuel, and the miscellaneous components required for the airplane's operation, which prevent the selection of the ideal shape purely from lift and drag considerations.

When the orientation of the vehicle is changed with respect to the flight direction, it presents a different shape to the oncoming airflow, and thus experiences a change in the aerodynamic forces and moments. The orientation of an object of an arbitrary shape can be described by a maximum of three angles relative to the specified flight direction. However, if the object has a plane of symmetry (which almost all flight vehicles have) then it is necessary to use only two angles to represent its orientation relative to the flight direction. As discussed later, these two angles are the *angle-of-attack* and the angle of *sideslip*. The sideslip angle gives the relative velocity component normal to the plane of symmetry, whereas the angle-of-attack is used to resolve the flow into two mutually perpendicular directions (forward and upward) in the plane of symmetry.

The atmospheric properties governing the lift, drag, sideforce, and the aerodynamic moments are the density, the pressure (or temperature), and the viscosity of the air prevailing at the flight altitude. Everything else remaining the same, the aerodynamic forces and moments are directly proportional to the atmospheric density. The atmospheric properties are modeled by assuming a standard variation with the altitude, which is briefly explained in Appendix.

For a given shape and size of the object, its orientation relative to the freestream, and for a specified set of atmospheric properties, the aerodynamic forces and moments are directly proportional to the square of the flight speed.

The reference length and area indicating an object's size determine the magnitude of the aerodynamic forces and moments experienced by it. The reference area is usually selected to be the wing's *planform area* defined as the area of the wing as seen from the top. The reference length is usually a characteristic dimension of the object along the expected flow direction. This length is called the *characteristic length*. For example, the characteristic length of an airplane wing's cross section at any spanwise location taken parallel to the freestream is its *chord*, which is the straight line joining the foremost point (the *leading edge*) to its rear-most part (the *trailing edge*). For the whole wing, the characteristic length is selected to be either an average (mean) chord of the wing, or its *span* measured from tip to tip.

The aerodynamic forces are often rendered non-dimensional by dividing by the product of the density of the freestream, the square of the freestream speed, and the reference area. Usually a factor of $1/2$ is also used in the division, because it produces the *dynamic pressure* when multiplying the density and the square of the airspeed. The dynamic pressure has a special place in incompressible flows. Hence, the non-dimensional aerodynamic force is derived by dividing the force by the product of the freestream dynamic pressure and a reference area. Similarly, a non-dimensional aerodynamic moment is obtained through the division by the product of the freestream dynamic pressure, the reference area, and the reference length. The non-dimensional forces and moments are called the *coefficients* of the respective forces and moments, and are independent of the size of the object, and the freestream conditions. They are very useful when determining the aerodynamic characteristics of an object by its scaled model in a wind-tunnel test.

2.2.1 Viscous Flow

The stickiness of a fluid is called its *viscosity*. A governing parameter for viscous flows is the *Reynolds number*, which is defined as the ratio of the flow's momentum per unit cross-section area to its viscosity. Thus the Reynolds number of a flow is related to the shear stress created by the flow. A highly viscous fluid such as the molasses flowing over an object at a very low speed has a small Reynolds number and hence large shear stress, whereas a gas flowing rapidly over the same object would have a high Reynolds number and a small shear stress. The Reynolds number is directly proportional to the density and the speed of the flow, but inversely proportional to its viscosity. For example, water is 1000 times denser, but only about 100 times more viscous than the air. Hence, the Reynolds number of an object moving with the same speed in water is about ten times higher than that in air. For a given speed, density, and viscosity, the Reynolds number is proportional to the characteristic length. The Reynolds number based upon the mean chord of a typical airliner while cruising is of the order of ten million, whereas that of a small insect can be only 100.

2.2.2 Streamlined Shapes

Since air is a gas with a low viscosity, the magnitude of shear stresses on the wing of an airplane is typically much smaller than the exchange of momentum normal to the flow (pressure), because of its high Reynolds number. Therefore, the shape of an airplane wing can be designed to produce the maximum possible lift due to the pressure variations over it. However, even a small viscous shear stress results in a drag that must be balanced by the thrust of the engines, and determines the fuel consumption (hence flight efficiency). The streamlined shape is derived from the technical term *streamline*, which is defined as a curve in a steady flow whose tangent at any point gives the local flow direction. The streamlines help us in discussing what eventually happens to a steady flow which is uniform far upstream of an object such as a wing. The pattern of a steady flow can be shown in a diagram by a set of streamlines passing around the object under study. The uniform flow far upstream of the wing is depicted (as in Fig. 2.1) by parallel straight lines. When passing a smooth solid surface such as the wing, the streamlines must be parallel to the local surface, because the flow cannot go through it.

When the streamlines faithfully follow the external contours of the flight object, there is the maximum possible exchange of momentum between the flow and the solid object. Such a flow is called an *attached flow*. By designing the vehicle with flat and thin surfaces like the wings and tails, and slender shapes like the fuselage and nacelles, it is possible to keep the flow largely attached when essentially flying along the longer vehicle dimension (the *longitudinal axis*). Such a shape is referred to as a *streamlined shape*, and the generally small angle made by the longitudinal axis with the flight direction is called the *angle-of-attack*. For a wing, the longitudinal axis is the chord, while that of a fuselage or a nacelle is its center-line dimension. The pressure variations on the wings and tails then give rise to a force normal to the flight direction, which is called the *lift*. This is the beneficial part of the aerodynamic effects, and enables heavier-than-air flight. By increasing the angle-of-attack at a given speed, it is possible to increase the magnitude of the lift. However, there is a limit on the extent up to which lift can be increased with the angle-of-attack.

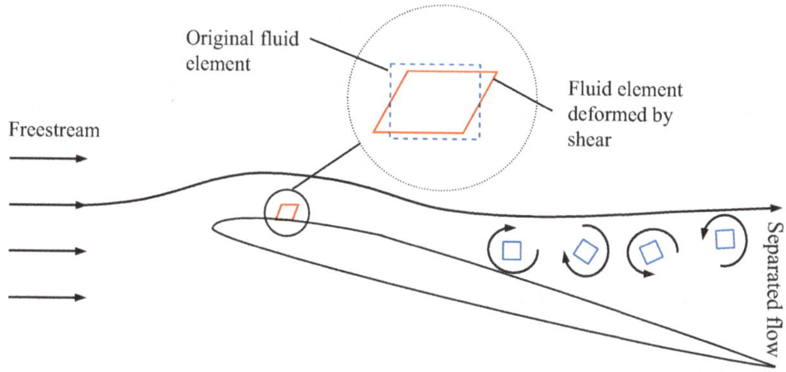

Fig. 2.1 A steady and uniform freestream flow passing a streamlined solid object (a wing) can encounter separation from the object due to a steady loss of momentum of fluid elements caused by skin friction. The shear deformation and eventual rotation of the fluid elements sap energy from the flow, thereby creating the skin friction drag and the pressure (or form) drag, respectively

Unfortunately, even the most well-designed streamlined shape experience an adverse phenomenon caused by the pressure variations along the body surface. This effect called the *pressure drag* (or form drag) is due to the continuously decelerating flow near the surface caused by the skin friction. As a fluid element adjacent to the body decelerates, it eventually reaches a point where its speed has fallen below that required to provide the necessary centripetal acceleration for hugging a convex part of the body. (Recall from Chap. 1 that the centripetal acceleration around a curve at a constant speed is proportional to the square of the speed.) Therefore, a streamline is no longer able to follow a convex contour, and is said to *separate* from the body, which is depicted in Fig. 2.1 as an example of the flow past a wing at a large angle-of-attack. The separation of the flow usually happens near the aft part of a streamlined shape, but the point of separation can move upstream if the body presents a higher inclination (angle-of-attack) to the flight direction. As the flow completely separates from the body, it is no longer able to exchange any momentum with the body, and hence the magnitude of pressure variation (thus the lift) caused by the flow falls abruptly. The steep fall in the lift due to the separated flow is accompanied by an increase in the drag. This is the *stall* condition experienced by all wing-like shapes at a large angle-of-attack.

In order to understand how the pressure drag is created, let us focus on the fluid element depicted in Fig. 2.1, the lower part of which was initially in contact with the body and thus stationary with respect to it. However, the upper part is unrestrained by friction, thus an internal stress is created in the element which tends to deform its shape by a mechanism called *shear*. Energy is lost in this process, which results in the skin friction drag. The situation is similar to a piece of rubber whose lower extreme is at rest, but the upper extremity is being moved tangentially, causing it to deform as shown in the figure. As soon as the element separates from the body, its lower part is no longer being restrained by a contact with the body. In the process, its internal shear stress is relieved and its shape springs back to the original one. However, the element now starts to rotate in response to the moment created by the different forces applied at the top and the bottom. This is like a person slipping on ice and tipping over, because the ice cannot provide the necessary tangential force to balance the motion of the upper body. A similar slowing down and eventual separation takes place in all the layers of fluid adjacent to the layer which actual comes into contact with the surface. The rotary motion (called the *vorticity*) of the separated fluid elements takes away a large part of tangential momentum from the flow, which appears as an increase in the net drag on the body called the pressure drag.

The explanation given here is a simple way of understanding an important phenomenon, where the fluid elements are treated as if they were solid bodies. Actually, fluid mechanics is governed by the considerations of mass and energy conservation, in addition to the momentum conservation (Newton's laws) applied not to individual particles, but to a large collection of them. A detailed computation of the aerodynamic forces and moments thus requires the solution of the mathematical equations governing fluid mechanics.

2.3 How Is Lift Created?

A simple way to understand the creation of lift is by considering the flow past an airplane wing. Suppose the airplane is moving at a constant speed in a straight line relative to a stationary atmosphere. Since the relative motion takes place at a constant speed in a fixed direction, the air flow experienced by the wing is steady and uniform at every point located far upstream of the wing, as shown in Fig. 2.2. For simplicity, we also assume the wing has a constant chord and an infinite span. In such a case the flow cannot go around the wing tips, hence it is forced to remain in the two-dimensional plane formed by the page. In that case, the aerodynamic characteristics are the same everywhere along the span because there is no spanwise component of the flow. Such a flow is called a *two-dimensional flow*, and is depicted in Fig. 2.2. The cross section of the wing taken parallel to the freestream (thus

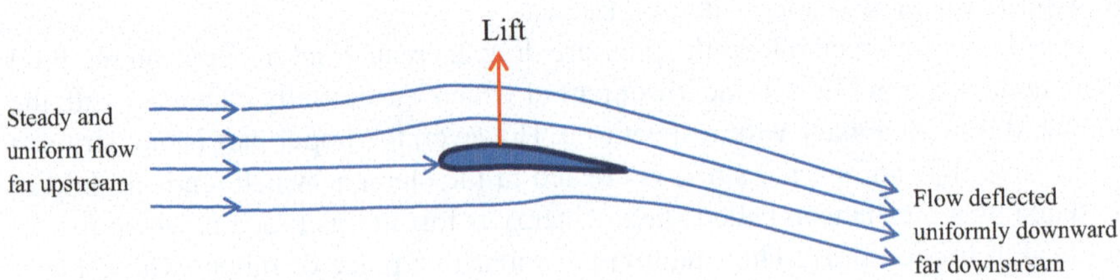

Fig. 2.2 A steady and uniform freestream is deflected downward by a positively cambered airfoil, and hence experiences an upward lift force that is equal and opposite to the rate of change of momentum of the flow

normal to the span) is called an *airfoil*. As shown in Fig. 2.2, an airfoil has a particular shape, which is essentially the same throughout the span. Its front part (the leading edge) is rounded, the top part is curved downward, whereas the bottom part is relatively flatter, and the rear portion ends in a sharp point (the trailing edge). Such a cross-sectional shape of the wing is said to have a *positively cambered* shape. If both the upper and the lower surfaces are symmetrically curved about the chord plane, the airfoil is said to be *symmetrical*. In a rare case, one could have a *negatively cambered* airfoil in which the lower surface is curved *upward* more than the upper surface.

2.3.1 Downwash

For simplicity, we assume that the flow is entirely subsonic, i.e., its speed everywhere is less than the speed of sound. In a subsonic flow, the pressure disturbance caused by the wing can travel upstream, thus the flow can begin to deviate from uniform flow far upstream in such a way as to accommodate the wing. This is evident in Fig. 2.2, where the steady streamlines begin to diverge much before reaching the wing. If this were not the case (i.e., if the flow were supersonic), the flow would experience a "shock" on arriving at the leading edge of the wing. We will discuss supersonic flow a little later. Since the subsonic streamlines smoothly deviate so as to follow the local curvature of both the upper and the lower surfaces of the airfoil, a positively cambered airfoil will produce a downward deflection of the flow as shown in Fig. 2.2.

The net change in the flow direction caused by merely passing the wing can be resolved in two mutually perpendicular directions, as shown by the vector triangles of Fig. 2.3. Let us first take the ideal situation when there is no friction (viscosity) in the fluid medium; thus there is no change in the magnitude of the flow's velocity

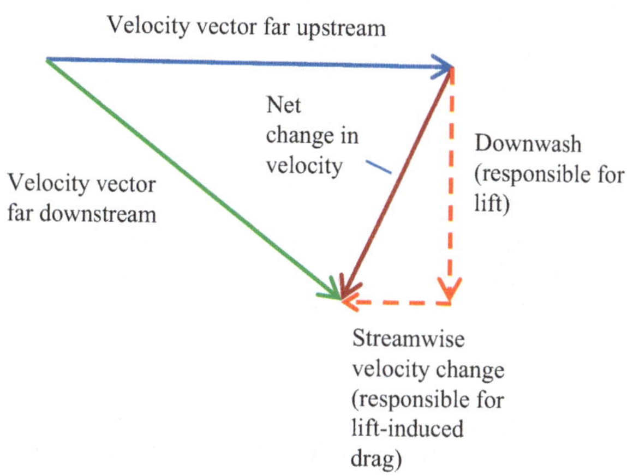

Fig. 2.3 Vector diagram showing lift generation by downward rotation of the flow's velocity vector due to the wing. Even if there are no losses due to skin friction and flow separation—thereby implying the same flow speed both far upstream and downstream of the wing—there is an attendant decrease of linear momentum due to the downward rotation of the flow's velocity vector. This is the lift-induced drag

(i.e., the relative speed) in passing the wing. It is depicted in Fig. 2.3 by the dashed arrows that even in this "ideal" case of inviscid flow, there are both a downward change (downwash) and a forward change (i.e., its decrease in the freestream direction) of the flow's velocity due to the wing.

The downwash created by the wing causes a net change of momentum of the flow in the downward direction. Because a fluid particle passing the wing experiences a downward (normal) force causing a change in its direction, it applies an equal and opposite reaction to the wing in the upward direction, normal to the freestream by Newton's third law of motion (see Chap. 1). The summation of the forces normal to the freestream exerted by all the fluid particles is the net lift force on the wing. The magnitude of the lift can be calculated from Newton's second law of motion by considering the net rate of change of momentum of a mass of fluid passing the wing at any given time, in the direction normal to the freestream.

A careful analysis of the vector diagram describing the flow's velocity components in Fig. 2.3 reveals that the velocity component along the freestream direction has decreased in the magnitude after passing the wing. This change in the flow's momentum in the forward direction (i.e., against the freestream) creates a net drag—a force in the backward direction (i.e., along the freestream)—by the Newton's third law. This force is called the *lift-induced drag*, because it is produced by the very mechanism that generates lift, namely, by rotating the flow's velocity vector in the downward direction.

When the fluid's viscosity is also taken into account, the net drag is much larger than the lift-induced drag, because a decrease in the flow's velocity takes place due to friction. A fluid particle in the neighborhood of the wing is affected by the friction caused by the wing's surface, and slows down having passed it, resulting in a smaller velocity of the flow far downstream as compared to that far upstream. This loss of fluid's momentum results in both the velocity components of Fig. 2.3 decreasing in their magnitudes, which increases the drag and slightly decreases the lift. The decrease in the lift happens because the downwash has a smaller magnitude due to viscosity. Consequently, viscous effects are detrimental to flight as they decrease the aerodynamic efficiency of a wing, which is measured by the lift-to-drag ratio.

2.3.2 Pitching Moment

Apart from producing the lift and drag, the steady flow past a wing also generates a *pitching moment* about any arbitrary point, O. The reason for this is once again the net rotation of the flow in the downward direction normal to the freestream. The generation of a rotational flow component (called the angular velocity) which induces a downwash far downstream of the wing, causes a rate of change of the flow's angular momentum in a clockwise direction as we look into the plane of Fig. 2.2. This rate of change of angular momentum of the flow creates a net torque about the wing in the opposite (i.e., counter-clockwise) direction by Newton's second and third laws, which is the pitching moment. The pitching moment thus created depends upon the magnitude of the downwash, therefore also on the lift being produced by the wing. If the lift is higher, there is a greater downwash associated with it, and hence the pitching moment is also greater. The variation of the lift, drag, and pitching moment produced by the airfoil in the various flow conditions gives a complete description of its aerodynamic characteristics.

The pitching moment created by the wing depends upon the location of the point, O, about which the moment is taken, as well as on the lift. The effect of the drag on the pitching moment is negligible due to its much smaller magnitude, as well as a much smaller moment arm for an airfoil compared to those associated with the lift. The dependence of the pitching moment on the lift can be understood by considering that the lift is independent of the arbitrary location of the point O. However, the contribution of the lift to the net pitching moment is different about each location of O, due to a change in the moment arm of the lift from that point. Reasoning in this manner, we can surmize that there must be a location of the point O about which the net pitching moment is independent of the lift, its moment arm being zero at that point. Such a point is called the *aerodynamic center*, and its location is such that there is no variation in the net pitching moment about it, even if there is a change in the downwash (thus the lift). Since the lift for a given freestream properties (airspeed, density, etc.) can change only if there is a change in the angle-of-attack, the pitching moment about the aerodynamic center is also independent of the angle-of-attack. For a thin airfoil in a subsonic flow, the aerodynamic center is located near the quarter chord point measured from the leading edge, and moves near the mid-chord location in a supersonic flow.

The presence of the aerodynamic center implies that in the zero-lift condition, the airfoil creates a pure couple resulting in the pitching moment having exactly the *same* value about any arbitrary point, O. For a positively cambered airfoil, the zero-lift pitching moment tends to lower the nose of the aircraft. This is called a nose-down pitching moment. Since a nose-up pitching moment is considered to be positive by convention, the pitching moment produced by a positively cambered airfoil is negative. A symmetrical airfoil does not have a zero-lift pitching moment, whereas a negatively cambered airfoil has a positive (nose-up) value of the zero-lift pitching moment. Figure 2.4 depicts the lift, drag, and the zero-lift pitching moment on a cambered airfoil.

The location of the aerodynamic center and the value of the zero-lift pitching moment are important aerodynamic parameters when the aircraft's stability and control are considered. We will return to them in Chap. 3.

2.3.3 Pressure Distribution

In the ideal case of an inviscid flow, the lift is entirely created by the variation of the local pressure exerted by the flow over the wing. Since the pressure is the force per unit area, the summation of the forces acting on the lower surface normal to the freestream direction must exceed that on the upper surface for a positive lift.

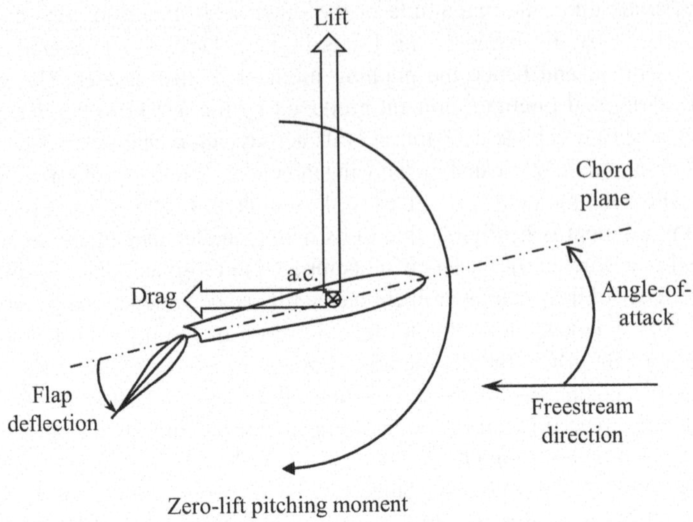

Fig. 2.4 Lift, drag, and zero-lift pitching moment about the aerodynamic center (a.c.) of a positively cambered wing at an angle-of-attack

Clearly, the lift, the net difference in the pressure distribution between the lower and the upper surface, and the downwash created by the wing are mutually related. An increase in any one of these quantities causes an increase in the other two. For example, an airfoil with a larger camber will generate a larger downwash, thereby increasing both the pressure difference and the lift for the same flow conditions (freestream speed, density, etc.). This is the very mechanism of increasing the lift with the help of movable flaps during take-off and landing. The deflection of leading-edge and trailing-edge flaps increases the effective camber of the airfoil, which produces a larger downwash, and thus a greater lift. The example of a trailing-edge flap is shown in Fig. 2.4, whose deflection increases the lift, the drag, and the zero-lift pitching moment due to an increase in the effective camber of the airfoil. A similar effect is obtained by a leading-edge flap. Large airplanes such as the airliners, and high-speed fighter type airplanes employ both leading-edge and trailing-edge flaps. The increased lift due to the deflection of the flaps helps while taking-off and landing, because it decreases the airspeed required for flight. A similar increase in the lift also helps during maneuvering. In the normal cruising condition, the flaps are retracted, because the same lift can be now provided by the increased flight speed with a smaller effective camber of the airfoil.

The lift and the pitching moment of an aircraft are calculated from the pressure distribution on the wing and the tail. The viscous effects modify the pressure distribution computed by assuming an inviscid flow. Such a change is usually small at the small angles-of-attack required for the normal cruising flight of an aircraft, and can thus be neglected. Therefore, the lift and pitching moment in most cases can be estimated from an inviscid pressure distribution. However, at the larger angle-of-attack required during take-off, landing, and slow maneuvers (see Chap. 3), the viscous effects can cause an extensive flow separation (as previously explained), and must be taken into account while calculating the pressure distribution.

2.4 Vorticity and Circulation

Let us consider what happens to the flow in the two-dimensional case when it is deflected downward by the wing. We can appreciate this by the example of a skier, who was traveling in a straight line at a constant speed before being turned by a ski pole briefly stuck in the snow. The resulting force applied normal to the direction of motion causes a turn in the direction of the force. The skier briefly describes a curve as long as the force is being applied, but returns to a straight-line motion (in a new direction) as soon as the pole is lifted from the snow and the force is removed. By a similar process, the fluid elements passing a wing are turned downward as shown in Fig. 2.2. In the process of being turned by the wing, the elements describe curved paths depicted by the streamlines in the wing's vicinity.

Returning to the analogy of the skier, when the pole is used to apply a force normal to the direction of motion, a torque is also applied on the skier's body. This happens because the pole hits the snow a little away from the body of the skier. The torque would cause the skier to trip and rotate about a vertical axis passing through her center of mass, unless another torque is quickly applied by the other pole in the opposite direction.

A fluid element is like a skier with only one pole, because it can interact with the solid body (wing) on only one of its sides. In addition to the force applied to deflect the particle's trajectory in the normal direction, a tripping action (torque) is also present, which causes the particle to start spinning about an axis passing through its center of mass (as indicated for the separated fluid element in Fig. 2.1). Therefore, the wing supplies both a downward velocity and an angular velocity to all the fluid elements in its vicinity. If the fluid is inviscid, there can be no opposing torque applied by the neighboring elements, and the fluid element continues to spin at a constant angular rate as it deflects downward. This is called an *irrotational flow* where the net vorticity (or angular velocity) of the flow is conserved.

By what exact mechanism can the rotation of an air element affect the lift of a wing in the attached flow situation? As discussed earlier, there are only two ways a fluid can exchange momentum with a solid surface: (a) a normal pressure and (b) a tangential shear stress exerted on the surface. By Newton's third law, the surface exerts equal and opposite forces on a contacting fluid element. If there is no viscosity in the fluid medium, the elements away from the solid surface can only be affected by the changes in the pressure transmitted to them by the intervening particles. However, all real fluids have some viscosity, thus a shear stress is also applied by one fluid particle to the other along the direction of motion. The magnitude of shear stress experienced by two contacting surfaces is proportional to the difference in the speed (relative speed) of the surfaces. When two adjacent fluid particles are moving at nearly the same speed, they experience a negligible shear stress. This is the situation in a typical air flow far away from the wing. But very close to the wing, we have the opposite situation. Here the viscous effects are dominant, because large velocity differences exist between adjacent layers of fluid. A fluid element is slowed down very abruptly in contact with the wing's surface, as if brakes were applied to it. This is due to a large shear stress on the contacting particle because of its large initial speed relative to the wing.

The velocity difference between the fluid layer that has almost come to rest after contacting the wing and an adjacent moving layer causes a shear stress on the moving layer. This shear stress is slightly smaller than what was experienced by the layer which contacted the wing. In this manner, a shear stress of a successively smaller magnitude is experienced by fluid layers farther and farther away from the wing. Due to the difference in the shear stress magnitudes on the vertically opposite sides of a fluid element, a torque is applied on it, which rapidly decreases in the magnitude as the vertical distance of the element from the wing increases. As seen earlier, this torque causes the elements in the vicinity of the wing to begin rotating, while those elements which are a little away from the wing do not experience any appreciable torque, and consequently do not rotate. This is the only mechanism of producing vorticity (or angular velocity) in the flow. In this discussion, we have arrived at the heart of an important idealization. Far away from the solid boundary, viscous effects of an air flow are negligible, therefore it can be treated in the same manner as if it were inviscid. But very close to a solid boundary, the inviscid flow approximation is invalid, and the rotation of the fluid elements under shear stresses must be considered.

It is now clear that lift generation cannot be completely understood unless the rotation of the fluid elements due to the wing is also described. Consider a closed ring formed of specific fluid elements moving with the flow as shown in Fig. 2.5. Far upstream of the wing, the ring is merely translating, i.e., all the elements constituting

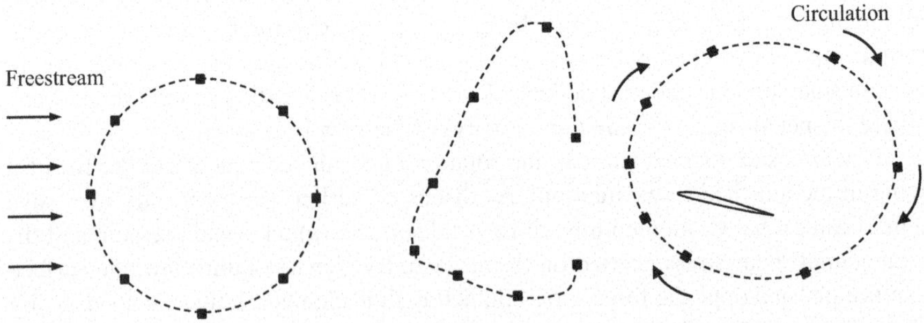

Fig. 2.5 Circulation in a closed curve formed by the same fluid elements around a wing airfoil is due to the net vorticity produced by the wing inside the closed curve. The flow circulation is thus responsible for the lift generation, as it is produced by the very mechanism that causes a rotation (vorticity) of the fluid elements. If there is no circulation of a ring of fluid elements enclosing the wing, the lift would be zero

it are moving together with the same velocity. If one were to sum up the flow velocity *along* (i.e., tangential to) the closed ring, then the result is called *circulation*. Due to the symmetrical flow pattern far upstream, the circulation is exactly zero there. As the wing is approached, the shape of the ring is distorted, because the fluid elements now have different velocities. In crossing the wing in the two-dimensional (infinite span) case, the ring must break up, because the fluid elements cannot travel in the spanwise direction to go around the wing.

Now consider an instant when the ring of the specific fluid elements is re-formed (shown in Fig. 2.5) and encloses the wing. The circulation in this case will have a non-zero value, because of the asymmetrical flow pattern around the wing. This circulation must have been supplied during the process of breaking-up and re-formation of the ring. Furthermore, the fluid elements enclosed by the ring are themselves rotating due to the viscous shear stress imparted by the wing, as discussed previously. Therefore, there is a relationship between the circulation around a closed curve and the net angular velocity (or vorticity) of the elements enclosed by the curve.

If there is no downwash produced by the wing, the lift is zero due to a symmetrical flow pattern around the wing, and hence the circulation vanishes along a closed curve enclosing the wing. Consequently, we have a direct relationship between the lift produced by a wing and the flow circulation around it.

Termed briefly, vorticity must be created by the wing in order to turn the flow downward. If there were no mechanism of producing vorticity (thus a circulation around a closed curve) by the wing, there would be no lift generated by it. The only mechanism of producing circulation in an attached flow condition is by the viscous interaction of the fluid layer adjacent to the wing. Due to the airfoil shape of the wing, just the right amount of circulation is created by it that leads to the lift without causing a flow separation.

The magnitude of circulation required to produce lift is practically calculated by assuming that the flow leaves tangentially to the airfoil's chord at the trailing edge. The basis for this assumption—called the *Kutta-Joukowski condition*—is the visualization of the flow as it leaves the airfoil. The physical process producing this condition is nothing but the viscous shear on the fluid layer adjacent to the wing. If there is no viscosity in the fluid, there can be no circulation, nor any "tripping" of the fluid elements to create vorticity for generating the downwash.

2.4.1 *Effects of Angle-of-Attack*

When examining the effect of the angle-of-attack on the lift, drag, and pitching moment on a given wing, we assume that the freestream conditions (the airspeed, density, temperature far upstream of the wing) remain constant. This implies that we are actually interested in the variation of the non-dimensional force and moment coefficients with the angle-of-attack. Once again we confine the discussion to the subsonic case, and only consider the wings of moderate to large aspect ratios for simplicity. The circulation, the downwash, the lift coefficient, and the pitching moment coefficient all increase linearly with the angle-of-attack in the attached flow condition. However, as the angle-of-attack is increased to a moderately large value, flow separation takes place, causing a decrease in all these quantities when compared to the attached flow situation prevailing at the smaller angle-of-attack. Hence, the variation of the lift and the zero-lift pitching moment coefficients become curves, rather than straight lines, reaching their respective maximum values at the *stalling* angle-of-attack as shown in Fig. 2.6. A further increase in the angle-of-attack causes an abrupt drop of the lift, and a variation in the zero-lift pitching moment due to a complete flow separation. This is called the stall condition. The stalling angle-of-attack thus determines the maximum lift. When a trailing-edge flap is deflected, the lift curve shifts nearly upward while maintaining the same linear slope (Fig. 2.6a). The lift at any angle-of-attack and the maximum lift coefficient are therefore increased; however, the stall condition is reached at a smaller angle-of-attack due to increased separation. The angle-of-attack corresponding to zero-lift becomes more negative as the flap deflection is increased. The zero-lift pitching moment coefficient also becomes more negative with the deflection of the trailing-edge flap. The variation of the pitching moment at the stall condition is an important flight characteristic of an airplane or glider, which will be discussed in Chap. 3.

The effect of leading-edge flaps (not considered here) is much more complicated, because it depends upon the wing's spanwise geometry, and the geometrical shape of the leading edge.

The non-dimensional drag coefficient varies linearly with the square of the lift coefficient for the attached flow case, producing a parabolic shape shown in Fig. 2.7. There exists a lift-coefficient, *A*, corresponding to the minimum drag, *B*. The value of the drag when the lift is zero is called the *parasite drag*. Figure 2.7 also shows the determination of the lift coefficient, *C*, corresponding to the maximum lift-to-drag ratio. This is obtained by drawing a tangent to the drag polar from the point *A*. The lift coefficient (and thus the angle-of-attack) for the maximum lift-to-drag ratio is an important performance parameter for an airplane or glider, which will be discussed in Chap. 3.

At higher angles-of-attack, the flow separation causes a departure from the parabolic shape of the drag polar, leading to a steep rise in the parasite drag (not shown in Fig. 2.7). In the stall condition, the lift being nearly zero does not produce any change in the drag, and a maximum possible, constant value of the parasite drag coefficient is thus obtained.

2.4.2 *Finite-Wing Effect*

The previous discussion was limited to the two-dimensional flow caused by a wing of infinite span, which did not allow a variation of the flow properties along the span. However, since all airplanes have only a finite span, the flow in the spanwise direction is also important. Such a flow is said to be *three-dimensional* in nature.

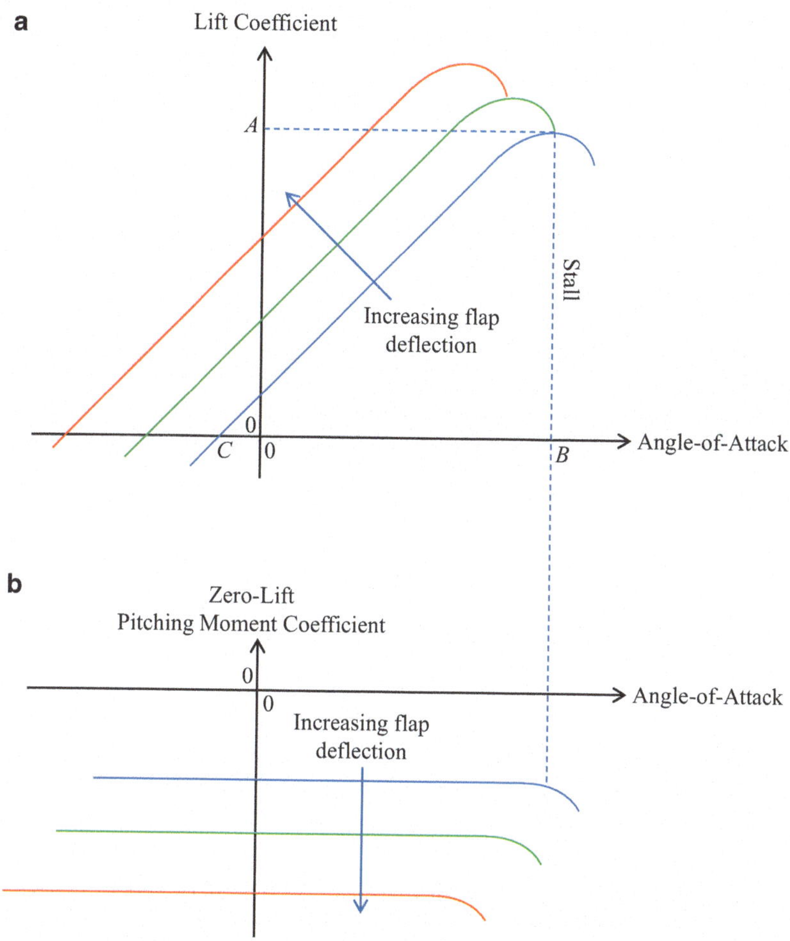

Fig. 2.6 The variation of the non-dimensional (**a**) lift coefficient, and (**b**) zero-lift pitching moment coefficient with the angle-of-attack for a positively cambered wing equipped with a trailing-edge flap. The stalling angle-of-attack for a given flap deflection (denoted by the point *B*) corresponds to the maximum lift coefficient, *A*, and decreases with an increased flap deflection. The zero-lift angle-of-attack (*C*) also varies with flap deflection, as shown. The zero-lift pitching moment is negative for a positively cambered airfoil, and constant with the angle-of-attack until the stall condition is reached

The effect of finite span can be understood by examining what takes place near the wing tips as the wing generates its lift. The pressure being larger on the lower surface than that on the upper surface of the wing creates a flow from the bottom to the top around the wing tips, as shown in Fig. 2.8. This flow must necessarily be rotating due to its very geometry of curving around the tips. Recall from Chap. 1 that a centripetal acceleration—thus a rotation—of a fluid element is required every time it has to go around a curve. Therefore, strong vortices are created at each wing

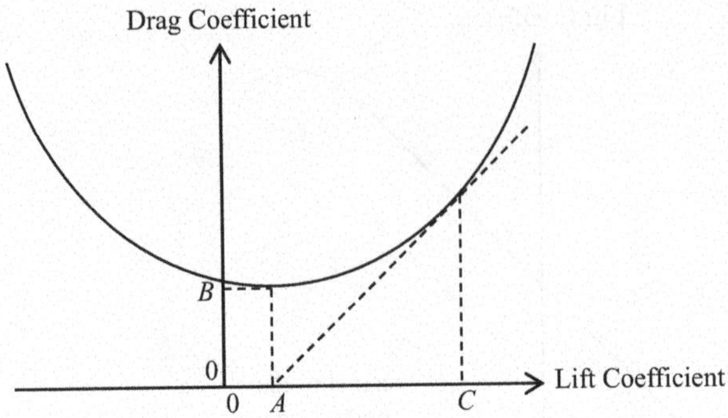

Fig. 2.7 The variation of the non-dimensional drag coefficient with the lift coefficient for a positively cambered wing. The minimum drag coefficient, B, is obtained for the lift coefficient at A, while the maximum lift-to-drag ratio occurs corresponds to the lift coefficient at C

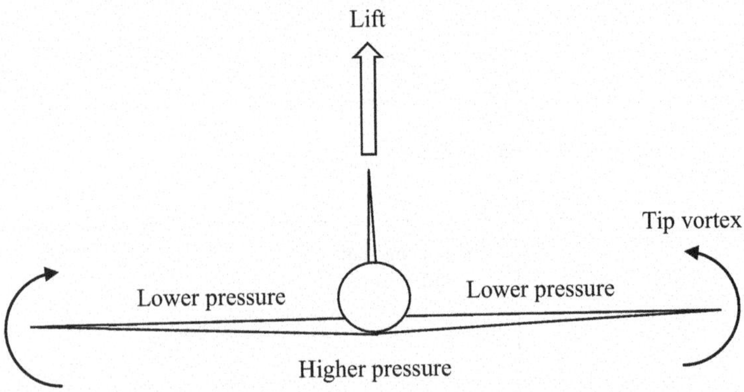

Fig. 2.8 Looking from the rear of an airplane or glider, the effect of a finite-wing span is seen to produce a circulation around each wing tip due to the difference in the pressure acting on the lower and upper surfaces of the wing. This results in wing-tip vortices which impart a normal flow component (downwash) at all points on the wing, thus decreasing its lift, and increasing the lift-induced drag

tip, causing a downwash at every point on the wing as indicated in Fig. 2.8. Due to the leakage of the flow around the tips, the pressure difference between the upper and lower wing surfaces, and hence the lift are naturally decreased. Furthermore, the downwash induced by the wing-tip vortices creates an additional lift-induced drag by the rotation of the flow's velocity vector (see Fig. 2.3). The net effects of a finite-wing are thus a decrease of the lift and an increase of the lift-induced drag from their respective values in the infinite span (i.e., two-dimensional flow case).

For a given planform area and flow conditions, a wing of a larger span has a higher lift, and a lower drag than a wing of smaller span. Thus the aerodynamic efficiency indicated by the maximum lift-to-drag (L/D) ratio increases with the span for a given planform area. The reason why the gliders tend to have the highest values of the wing *aspect ratio* (the square of span divided by the planform area) is that their range depends only upon the maximum lift-to-drag ratio. However, since a higher aspect-ratio wing achieves its maximum L/D ratio at a much higher angle-of-attack than one with a smaller aspect ratio, the airspeed corresponding to the maximum L/D condition becomes smaller. Consequently, an optimum combination of the aerodynamic efficiency and speed are obtained for most airplanes at moderate aspect ratios (less than 10), rather than about 20 employed in the high-performance gliders.

When operating close to the ground during the take-off and landing, a finite-wing's tip vortices become weaker in strength due to a partial obstruction of their flow from the ground. Therefore, the L/D ratio is increased appreciably from its value far away from the ground. This increase in the aerodynamic efficiency due to the ground is called the *ground effect*. The ground effect is appreciated by the pilots when trying to make a short landing, because it causes the flight path to stretch out due to the decreased drag and the increased lift. Due to its seemingly advantageous nature, the ground effect has also been employed in designing special airplanes (such as the Russian *Ekranoplan*) which can fly close to the ground for a better speed or economy. However, such vehicles can only operate over water for obvious reasons.

Chapter 3
Flight of Airplanes and Gliders: Vertical Plane

3.1 Introduction

Airplane and glider flight is enabled by aerodynamic lift, and essentially takes place in a straight line, punctuated by brief periods of maneuvering, which by definition, consists of curved flight paths. Except while making a horizontal turn, the flight of an airplane or a glider always takes place in the *vertical plane*, which is defined as the plane formed by any two radial lines emanating from the earth's center. That is why all airplanes and gliders are designed to have a *plane of symmetry*. By keeping the gravity vector in the plane of symmetry, all the forces can be kept naturally balanced in order to produce an equilibrium flight condition in a vertical plane. As soon as the plane of symmetry does not contain the gravity vector, an asymmetry exists in the aerodynamic forces and gravity, causing the vehicle to depart from the vertical plane in a horizontal maneuver.

Another feature of the airplanes and gliders is that their flight velocity is held constant for most of the flight, except during the brief periods of taking-off, landing, and sometimes also while maneuvering. This feature enables a simplified analysis of airplane and glider flights by assuming a steady flight condition.

A glider is an airplane without engines, whose design is optimized to produce a long gliding range and endurance. Any airplane becomes a glider when its engines are either switched off, or fail inadvertently. While gliders are presently used solely for recreation and sports, it is important to study the gliding characteristics of all airplanes in order to ensure a safe landing at an airport in an emergency resulting from engine failure.

In this chapter, we will examine the physical concepts behind the flight of airplanes and gliders in the vertical plane, whereas the next chapter is devoted to horizontal maneuvers.

A. Tewari, *Basic Flight Mechanics*, DOI 10.1007/978-3-319-30022-1_3

3.2 Straight and Level Flight

The most common flight condition of airplanes is the steady and level flight, where a constant altitude is maintained and the vehicle moves forward in a fixed direction at a nearly constant airspeed. The altitude referred to here is the height above the mean sea level, which is a standard reference point at the surface of the earth assumed to be a perfect sphere. Since the airplane follows the curvature of the nearly spherical earth by maintaining a constant altitude, its flight path between any two instants is really a circular arc, similar to the orbit of a satellite. However, since the variation in the angular position of the airplane takes place very slowly due to its relatively low speed when compared to that of a satellite, its trajectory in a steady and level flight can be approximated by a straight line. To a person on the airplane, the gravity appears to be always acting in a fixed downward direction. This is a familiar approximation that we apply to any object traveling on the ground, and is also the reason why the ancient people thought the earth was flat. The *flat-earth approximation* enables the trajectories on the ground as well as in the air to be analyzed by much simpler mathematics. Of course, the flat-earth approximation becomes inaccurate for vehicles approaching the orbital speeds, which is the case of most rockets and all spacecraft (to be discussed in Chap. 6).

We note here that the word "steady" applies to any situation in which the acceleration is zero. By assuming a straight-line flight at a constant speed, we are thus assuming a condition of zero acceleration. This is not exactly true because the airplane's horizontal trajectory is always curving slowly under the influence of gravity, which acts vertically toward the earth's center. Due to the curvature of the earth, the "horizontal" and "vertical" are not fixed directions (and planes), but keep on changing with the vehicle's position. For example, the vertical direction at the Empire State building in New York is very different from that at Qutub Minar in New Delhi, just because the two points are located far apart on the earth's surface. It is therefore logical to add the adjective "local" to the horizontal and vertical directions. What appears to us as the fixed vertical direction is only the local vertical. The plane normal to the local vertical axis is the local horizontal plane. We will return to this topic in Chap. 6.

It is thus clear that a centripetal acceleration must be present on any object traveling horizontally. Since the centripetal acceleration varies directly with the square of the speed, it is actually much smaller for an airplane—when compared to a launch vehicle or a spacecraft—and hence can be considered negligible in most cases. Consequently, the straight and level flight is an approximation. The phrase "straight and level flight" will be abbreviated as "level flight" in this chapter, because here we are focusing on the flight in a vertical plane. In the next chapter, which is for horizontal maneuvers, we shall return to the unabridged terminology.

Since an airplane flying level is traveling horizontally, it is necessary that the lift should be vertical and must also exactly balance the weight. Such an equilibrium condition is naturally obtained by maintaining a constant airspeed, because the lift increases with the square of the airspeed (see Chap. 2). A flight in which the speed is not constant would require the magnitude of the lift to be changed by either increasing or decreasing the angle-of-attack. However, any pilot will attest to the fact that it is difficult to perform changes in the angle-of-attack with precision. Therefore, a flight with a varying airspeed is generally not level, and the airplane tends to bounce up and down. A student pilot just beginning to learn flying, experiences such a situation, because the angle-of-attack is kept varying by the arbitrary application of the control inputs.

Because the natural equilibrium for level flight requires maintaining a constant flight speed at any given time, the thrust provided by the engines must exactly balance the drag of the airplane. Where this not so, the required level flight speed will not be maintained, and the airplane will either ascend or descend due to the changed lift. The equilibrium condition of steady and level flight in which the lift balances the weight, and the thrust balances the drag is depicted in Fig. 3.1. Here we are only considering the translation of the center of mass in describing the flight, rather than the complete airplane dynamics. The rotation of the vehicle about its center of mass has thus been excluded, and will be discussed later. For this reason, all the forces in Fig. 3.1 are shown as acting at the center of mass, which is depicted in the figure as a small circle with a "+" sign in the middle.

The steady and level flight is easily analyzed by energy considerations. Since the airplane is traveling at a constant airspeed and altitude, if there are no winds,

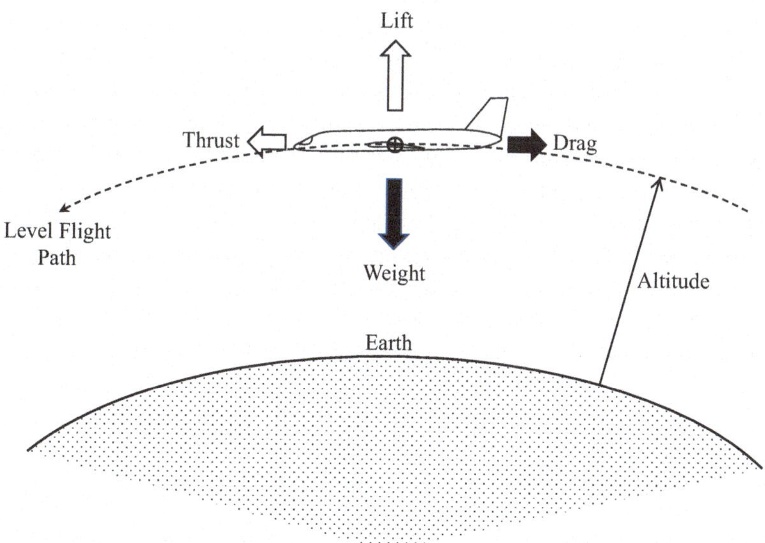

Fig. 3.1 The level flight of an airplane over the earth. The lift balances the weight and the thrust equals drag, thereby producing an equilibrium condition in which both the airspeed and the altitude are constant. The flight path naturally follows the earth's curvature, but appears to be straight to the passengers and crew because the flight speed is quite small. Note that the figure is not to scale (the airplane would be actually a dot on this scale), and only the translational motion of the center of mass (+) is considered here

its total energy is maintained a constant. However, since the drag always opposes the flight, its work done on the airplane is always negative and thus saps the total energy. The energy lost due to the drag must be exactly supplied by the engines, such that the net rate of change of energy with the time (i.e., power) is zero. Therefore, the power required for dragging the airplane through the air at a constant speed equals the power output by the engines. Airplane flight is thus a perennial tug-of-war between the thrust and the drag. If the drag exceeds the thrust, the balance of power is lost, and a descent results, provided the airspeed is maintained. The *gliding flight* is a descent in which the engine thrust is zero, therefore the vehicle is continuously losing energy (altitude). Conversely, if the thrust exceeds the drag required for steady and level flight, we have a *climbing flight* in which the altitude increases with time.

From the foregoing discussion, it is clear that a sufficient engine power must be available in order to balance the power required by the drag, D, at any given airspeed, v. The required power equals the product of the drag and the airspeed, Dv. If the net thrust produced by the engines is denoted T, then the available power is thrust times the airspeed, Tv. The balance of power therefore requires that the thrust must equal drag at the given airspeed. For an available engine power, the angle-of-attack can be adjusted by the pitch control (as discussed later in the chapter) in order to produce an airspeed at which the balance of power can be maintained. However, the airspeed at which the balance of power prevails must not become less than the stall speed; otherwise the airplane will stall (i.e., fall out of the sky) instead of maintaining a level flight. The stall condition therefore imposes a limit on the smallest possible airspeed at any given altitude for a steady and level flight. Another limitation is on the maximum possible airspeed at any given altitude due to structural reasons. Since the aerodynamic loads on the airplane's structure increase with the square of the airspeed at a given altitude, there is the possibility of a structural failure when a certain speed is exceeded. When atmospheric turbulence is also present, the structural speed limit is further reduced from its calculated value in smooth air. Unfortunately, there is no police in the sky to enforce these speed limits, and their violation often results in a catastrophe. To assist the pilot, airspeed indicators are clearly marked to show the stall speed and the maximum speed limits (both in smooth and turbulent conditions) for every certified airplane.

3.2.1 Flight Envelope

Figure 3.2 shows the plot of the power required for maintaining a steady and level flight vs. the airspeed at a given weight and altitude for a specific airplane. Note the shape of the required power curve where the airspeed for the minimum power required is well defined. At any airspeed different (either higher or lower) from this minimum power speed, there is an increased requirement of power. The thrust power output (called the *available power*) of the engines is also plotted in the same figure, showing an intersection with the required power curve at two points. These two

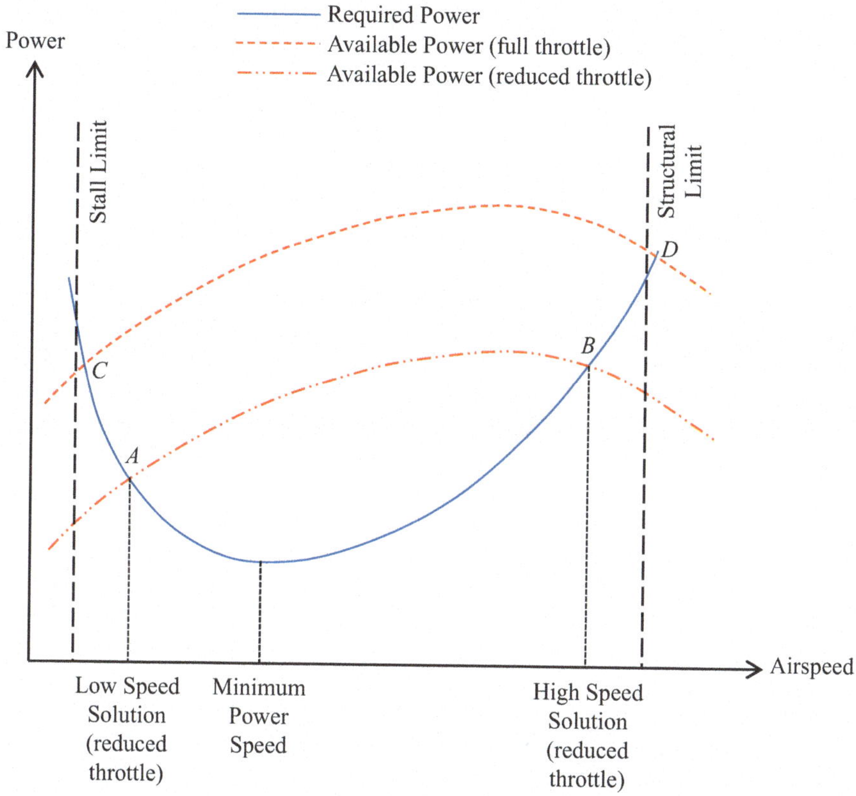

Fig. 3.2 Required and available power for flying steadily and level at a given altitude for a particular airplane of a given weight. The minimum power speed is determined from the shape of the required power curve. The balance of power at any throttle setting produces two airspeeds for steady and level flight—the low- and high-speed solutions, which are limited by the stall and the structural speed limits. If the altitude and the weight are changed, the required power, the available power, and the limiting speeds are also changed, resulting in new airspeeds for steady and level flight

points constitute the two possible conditions for achieving a steady and level flight at the given weight and altitude combination, and with the given power output. Which of these two conditions is actually met depends upon whether the corresponding airspeed falls between the stall speed and the maximum airspeed limits. By adjusting the engine power, the two points can be changed until the desired flight speed is produced. The value of the airspeed for level flight greater than the minimum power speed is called the high-speed solution, while that below the minimum power speed is termed the low-speed solution. If the low-speed solution is smaller than the stall speed, it cannot be a practical level flight speed. Similarly, if the high-speed solution falls above the maximum speed limit, it cannot be attained in practice.

Interestingly, the required power curve (Fig. 3.2) shows that when flying below the minimum power speed, an *increase* of power is required to achieve a *decreased* airspeed! This can be seen from the relative positions of the points marked A and C in the figure. The available power curve, AB, is for a reduced throttle setting of the engines, while the curve CD is for a higher throttle setting. As the engine power is increased, the high-speed solution increases in value from that at B to D, but the low-speed solution decreases in value from that at A to C. The behavior of the low-speed solution is to be contrasted with our everyday experience with the automobiles, where a decreased power always results in a smaller speed. This counter-intuitive nature of flying "behind the power-curve" (as termed by the pilots) is to be treated with caution, because of the possibility of stalling the airplane by trying to increase its speed with an increased power. Since low-speed flight is usually performed near the ground—either before the landing, or immediately after take-off—an inadvertent stall due to a poorly understood power curve could be dangerous.

Whenever either the low-speed or the high-speed solution lies outside the speed limits, it cannot be used as a practical flying speed. For example, the airspeed corresponding to point D in Fig. 3.2 is seen to exceed the structural speed limit. If the pilot is interested in flying at the high-speed solution, the engines must be throttled down such that the structural limit is not exceeded, and the point marked D either falls on the structural speed limit, or at a speed below it. Similarly, if the low-speed solution is smaller than the stall speed limit, the engines must be throttled down in order to fly at the low-speed solution.

The engine characteristics determine the available power curve at any given altitude, whereas the required power curve at the given altitude is determined by the aerodynamic characteristics and the weight of the airplane. If the altitude is varied for a particular airplane of a given weight, it is possible to plot a variation of the two speed solutions with the altitude. Such a plot, called the *flight envelope*, is depicted in Fig. 3.3 for an airplane of power characteristics shown in Fig. 3.2, flying at a specified weight. It is constructed by plotting the two speed solutions at full throttle, as well as the stall and structural limiting speeds, vs. the altitude. Since the flight envelope quickly informs us about an airplane's flight capability, it is an important tool in analyzing the airplane's performance. The steady and level flight is possible for an airspeed and altitude pair located inside (and on the boundary of) the flight envelope, but not outside it. The solid boundary $ABCDE$ in Fig. 3.3 is the flight envelope inside which a steady and level flight is possible without violating the stall and the structural limits. As the altitude increases, the available power decreases, and hence the envelope becomes narrower in its speed range. The maximum altitude at which steady and level flight can take place for an airplane with a given weight is called the *absolute ceiling*, and can be easily determined from the flight envelope. In Fig. 3.3, the point C denotes the absolute ceiling and the point D is the maximum airspeed possible in a steady and level flight for the airplane of a given weight. The maximum level flight speed at the mean sea level (point E) is smaller than that at D due to the structural limit. In this example, the structural limit is derived by imposing a restriction on the flight dynamic pressure (see Chap. 2).

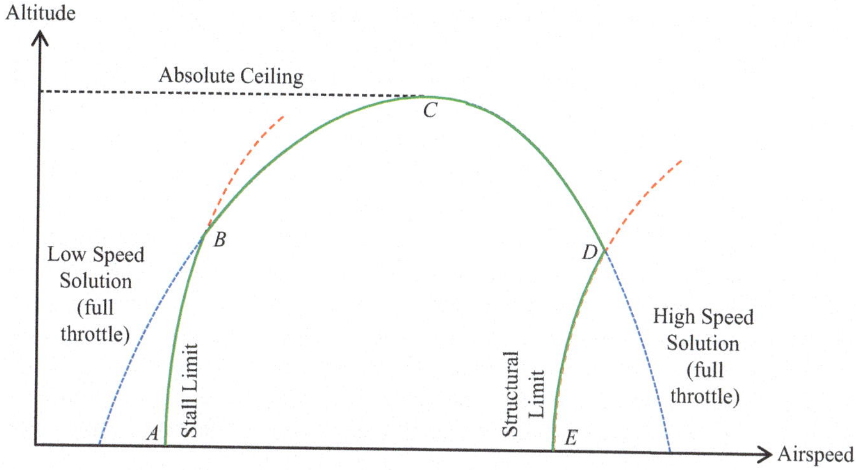

Fig. 3.3 The flight envelope of a particular airplane of a specific weight. It is the region of altitude vs. airspeed inside which a steady and level flight is possible. The boundary (*solid line*) of the envelope indicates the limiting case of either the full available power required for maintaining a steady and level flight, or the flight at the stall and structural limits. The maximum altitude of the flight envelope is the absolute ceiling

The curves shown in Figs. 3.2 and 3.3 are typical for a low subsonic airplane where the aerodynamic characteristics of the airplane remain fixed with the Mach number. As the speed of sound is approached and crossed, there is a large variation in the aerodynamic characteristics in a small range of Mach numbers due to the formation of shock waves. Consequently, the simple analysis presented in Fig. 3.2 is not possible for high subsonic, transonic, and supersonic airplanes. For example, the required power curve has a different shape at different Mach numbers, therefore it is no longer possible to talk of the minimum power speed and the low- and high-speed solutions. Since airplanes are powered by airbreathing engines (either propeller or jet), the available power is also governed by the freestream Mach number at which compressibility effects on the propeller, or at the jet engine's inlet can dictate the thrust developed by the engine. Furthermore, the stall and the structural limits are also functions of the Mach number. Such a dependence on the Mach number causes the flight envelope to take on a complicated shape, which could have many more corners than that of Fig. 3.3.

3.3 Cruising Flight

The flight in a steady and level condition in a fixed direction (heading) for an extended period is termed *cruise*. The distance traveled along the ground during a cruise is called the *range*, whereas the total time aloft is called *endurance*. Both the range and the endurance depend upon the quantity of fuel on board, the initial weight of the airplane, the airspeed, and the altitude at which the cruise is performed. For calculating the range accurately, the presence of any winds along the flight path (their speed and direction) must be taken into account. In trans-continental and inter-continental flights of several thousand kilometers, the earth's rotation must also be accounted for by considering the Coriolis acceleration. If this is neglected, both the range and the intended destination would be in error.

The altitude for cruise is selected from the engine efficiency considerations. The engine efficiency is measured by the mass of fuel consumed per unit of thrust (or power) produced by the engine. This is termed the *thrust-specific* (or *power-specific*) fuel consumption. Each type of engine has a particular thrust- and power-specific fuel consumption characteristics, which are defined by their variation with the altitude and the airspeed (or Mach number). For example, a propeller airplane driven by a piston engine can have its maximum power-specific fuel consumption (PSFC) at the standard sea level when operating at full throttle, irrespective of the airspeed, and decreases with the altitude for a given throttle setting. In contrast, a jet engine's thrust-specific fuel consumption (TSFC) varies with both the throttle setting, altitude, and airspeed. A turbo-prop engine—regarded as a hybrid between a propeller and a jet engine—has its PSFC varying with the throttle setting and the altitude. Consequently, while the cruise airspeed is defined by the airframe (lift and drag) characteristics and the weight, the altitude must be chosen so as to minimize the specific fuel consumption of the engine. A suitable airspeed and altitude pair for cruise is then selected from the airplane's flight envelope, as well as the standard atmospheric model, which is briefly discussed in Appendix.

The normally aspirated (i.e., with intake open to the atmosphere) piston engines do not have a minimum PSFC at any altitude. However, since their PSFC steadily decreases with an increasing altitude, the cruise altitude is selected by matching the power required for flying at an optimum airframe speed. The cruising altitudes of normally aspirated piston engined airplanes are limited to about 5 km in the standard atmosphere. Turbo-charged piston engines can maintain a constant power up to a certain altitude, and hence have a much higher cruise altitude (about 8 km in standard atmosphere). Jet airplanes have their minimum TSFC near the *tropopause* altitude (see Appendix), which is near 11 km standard altitude. Turbo-prop engines deliver their minimum PSFC at standard altitude of nearly 9 km.

The cruising airspeed is selected for the optimum aerodynamic efficiency, which yields the smallest fuel consumption for a given cruise altitude, initial weight, and either the range, or the endurance. This implies that either the largest range, or endurance would be obtained by flying at a particular optimum airspeed for a given altitude and weight. For every airplane, the airspeeds for best range and endurance are quite different. Propeller-driven airplanes deliver their maximum range when flying at the minimum drag condition for level flight. This is the airspeed at which the maximum lift-to-drag ratio, L/D, is achieved in a steady and level flight. In contrast, subsonic jet airplanes achieve their maximum possible range when flying at a airspeed, v, which maximizes the product of the airspeed and the lift-to-drag ratio, $v(L/D)$. However, because the required optimum airspeed for maximum range of subsonic jet airplanes usually crosses the subsonic regime, a smaller speed is more practical for structural reasons. This is of course at the cost of the range. Most subsonic jet airliners cruise at Mach numbers in the range 0.8–0.85, whereas their theoretically optimum Mach number would be much higher (around 0.92).

Beyond the subsonic regime, the formation of shock waves drastically decreases the L/D ratio, and increases the TSFC of the high-bypass turbofan engines employed in airliners. An airframe optimized for supersonic cruise—such as the erstwhile *Concorde*—must have a more slender shape of the fuselage, and a higher sweep angle of the wings in order to reduce the drag penalty associated with the shock waves. The engines for such an airplane must also be optimized for supersonic flight by a proper selection of the bypass ratio, and a suitable design of the air-intake and the nozzle. With the available technology, supersonic cruise is still beyond the reach of routine airline operation owing to a high TSFC and a low L/D ratio, costing as much as twice the flight of a subsonic airliner for a given range.

When the endurance is to be maximized, the flight must take place at a slower speed than that for the best range. This fact is intuitive, and is also based upon the shape of the required power curve (Fig. 3.2), where the minimum power required for level flight lies between the two extremes. For the maximum endurance, the rate of fuel consumption must be minimized. For a propeller engined airplane, the minimum rate of fuel consumption is achieved when flying at the minimum power airspeed corresponding to the cruising altitude. Recall that the PSFC is a function of the altitude for propeller airplanes, and can be reduced by selecting a proper altitude. If the power required at the selected altitude is also minimized with respect to the airspeed, then the smallest possible fuel consumption rate is obtained. For jet airplanes, the airspeed for the maximum endurance corresponds to that required for the minimum drag (maximum L/D) condition at the given weight. This speed is much higher (about 31 % more) than that required for the minimum power. The same speed advantage exists for jet airplanes in the maximum range operation. Consequently, an airplane equipped with jet engines flies much faster than the same airplane equipped with propeller engines for a given range or endurance. This is mainly the reason why nearly all airlines use jet engined airplanes for medium to long haul flights.

While an airplane is cruising, its weight is continuously decreasing due to the fuel being consumed. A typical long range flight can have an airplane at landing weighing only about 70 % of that at take-off. Since this weight loss happens over many hours, it goes unnoticed by the passengers. However, in order to maintain a cruising speed, the airplane is allowed to climb at a very small rate as its weight lessens throughout the cruise. Climbing slightly while cruising gives a small advantage in terms of the range (or endurance) when compare to a flight in which the altitude is strictly maintained constant. The resulting flight is thus not exactly level, and is referred to as a *cruise-climb*. Most pilots request the air-traffic control to be allowed a cruise-climb for fuel efficiency reasons. Often the cruise-climb is carried out in steps rather than continuously, as permitted by the air-traffic control.

3.4 Climbing Flight

Most airplanes must quickly reach their cruise altitude for the maximum possible range (or endurance). This requires them to execute a climb immediately after the take-off, taking them to the desired altitude. The most economical climb is the one in which the airspeed is maintained constant and the altitude is steadily increasing. This is because—as in the level flight—the ideal combination of the lift and drag must be achieved for the minimum required fuel consumption, and this requires an optimum airspeed (hence angle-of-attack) to be maintained. A climb takes place in the airplane's plane of symmetry, with the lift vector inclined with respect to the local vertical (see Fig. 3.4) at the climb angle. In a steady climb, the thrust must balance not only the drag, but also a component of the weight due to the inclination of the flight path. Consequently, the power required for climb is always higher than that required for a steady and level flight at a given altitude.

The rate-of-climb is the vertical speed component of the airplane during a climb, and is directly proportional to the excess of power available over that required to balance the drag alone. Maintaining the maximum rate-of-climb though a climb results in the smallest fuel consumption for a given increase in the altitude. Since propeller engines can deliver nearly the same power at different airspeeds, the maximum rate-of-climb for propeller engined airplanes is achieved by flying at the speed for the minimum required power. However, due to the flight path now

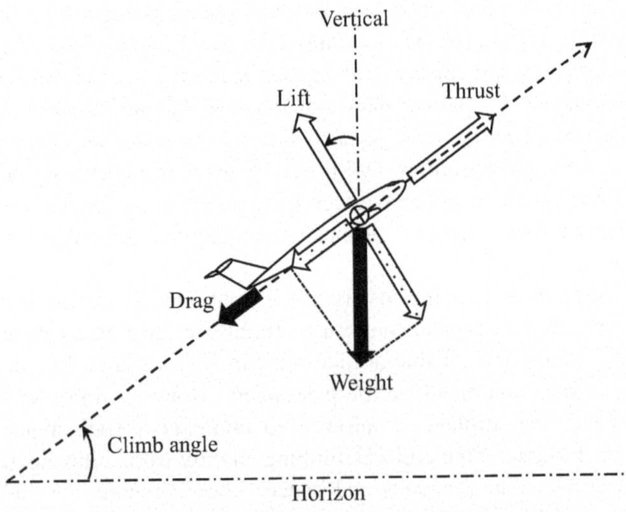

Fig. 3.4 Climbing flight at a constant climb angle and a steady airspeed. The thrust must balance the tangential component of the weight as well as the drag, whereas the lift only balances the component of the weight normal to the inclined flight path. Consequently, the engine power required for a climb is higher, and the airspeed lower, than the corresponding power and speed for a steady and level flight at a given weight and altitude

being inclined upward, the minimum power airspeed is smaller than that required for a steady and level flight at a given weight and altitude combination. The reason that the climbing airspeed is smaller than the corresponding level flight speed is that in a climb, the lift only balances a component of the weight (rather than the entire weight). The required smaller lift is therefore achieved at a smaller airspeed, everything else remaining the same. Jet airplanes achieve their maximum rate-of-climb at an airspeed which is neither the minimum power, nor the minimum drag speed.

Sometimes, it may be required to maximize the climb angle rather than the rate of climb. This may happen if a large obstacle (such as a mountain) has to be cleared. The airspeed required to maximize the climb angle is different for each type of airplane, as well as different from that for the maximum rate-of-climb. Some fighter type airplanes can have a vertical climb capacity (i.e., a climb angle of 90°) at the lower altitudes, because their maximum available thrust at such altitudes can exceed their weight. In such a vertical climb, the lift must be kept zero, otherwise the flight path will depart from being vertical. Due to the large excess power of their engines, the fighter type airplanes are also capable of accelerated climb, where both the airspeed and altitude are increased simultaneously.

In our discussion here, we have assumed both the flight path is a straight line and the airspeed is constant, which, if it were true, would result in a steady (non-accelerated) flight. However, if the airspeed is held strictly constant as the airplane climbs, an equilibrium cannot be maintained because the lift is a function of atmospheric density. As the altitude changes in a climb, so does the atmospheric density and the lift. Therefore, at every different altitude, a new airspeed is required for keeping the lift exactly balanced by the normal weight component. If this is not carried out, a departure from the straight-line climb will result due to the normal acceleration produced by the unbalanced lift. Hence, a "steady" climb is either carried out with a constant airspeed, resulting in a flight path curvature, or in a straight line, which requires a varying airspeed. If the rate-of-climb is small, the variation in the airspeed (or in the flight path) takes place very slowly, and hence the steady-state condition approximately exists at every time instant. Such a climb is called a *quasi-steady climb*. However, as remarked earlier, fighter type airplanes display an accelerated climb (also called *zoom climb*) capability due to their much larger power-to-weight ratio.

3.5 Descending and Gliding Flight

A descent from the cruise altitude is required for all airplanes before making a landing. Such a descent must be steady in order to be controlled by the pilot. Otherwise there is a possibility of the airplane entering a steep dive, which can lead to the airspeed exceeding the structural limit. The engines must be always throttled down from the steady and level flight, because an equilibrium in the descent requires a smaller power than that in the level flight. The picture presented in a steady descent

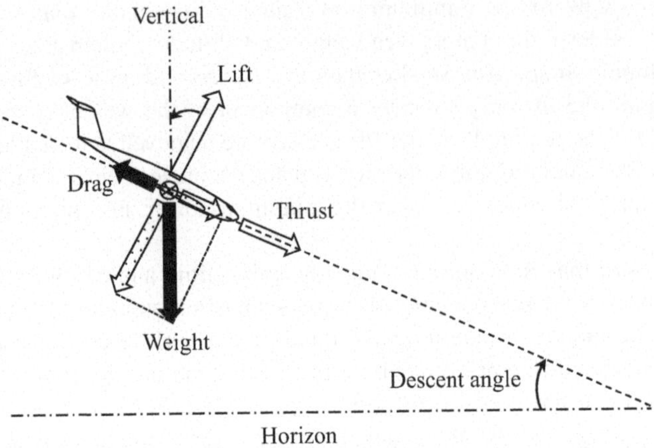

Fig. 3.5 Descending flight at a constant angle and a steady airspeed. The thrust is aided by the tangential component of the weight in balancing the drag, whereas the lift only balances the component of the weight normal to the inclined flight path. Consequently, both the engine power and the airspeed required for a steady descent are smaller than those needed for a steady and level flight at a given weight and altitude

(Fig. 3.5) is the reverse of that in a steady climb, where a component of the weight now aids the thrust in balancing the drag. Consequently, the descent is analyzed quite similarly to a climb, the only change being the replacement of a positive angle by a negative one for the flight path's inclination. However, the lift (and thus the airspeed) in a descent of a particular angle remain exactly the same as for a climb at the same angle.

We will now briefly consider the analysis of the gliding flight, which is a special case of the descent with a zero engine power. Since the component of the weight along the flight direction must balance the drag in a steady glide, the gliding slope is merely the reciprocal of the L/D ratio. Hence, the minimum glide angle—required for the maximum gliding range—is achieved by flying at the maximum lift-to-drag ratio (L/D) condition. Most gliders have a very high L/D ratio, often greater than 30; therefore their gliding slope is smaller than 1 in 30, implying a loss of 1 m of altitude for every 30 m traveled forward. Since the minimum glide angle is quite small (less than 2°), it is accurate to assume that the lift equals the weight, as in a steady and level flight. Hence, all the airspeed calculations can be performed merely by substituting the weight for the lift. The stall speed is thus the same as that in the level flight, and so is the minimum power airspeed. The maximum endurance for a glider is obtained when flying at the minimum power airspeed (Fig. 3.2), because it gives the smallest descent (or sink) rate. The speed required for the maximum gliding range is extremely important, even for airplane pilots, because it gives the best chance of reaching an airport in the event of a complete engine failure when all airplanes become gliders. Similarly, the maximum gliding endurance airspeed is important since it allows the maximum available time for a pilot to try to re-start a failed engine. If these speeds are not calculated in advance, there is a danger of encountering a stall while trying to achieve either the best gliding range, or endurance. Needless to say, a stall in this case could be disastrous, because it leads to a sudden and large loss of altitude. Remember: altitude is the "fuel" for a glider!

The remarks relating to a quasi-steady climb are also applicable for the steady descent. If the airspeed is held constant, the flight path cannot remain exactly straight. Conversely, if the flight path is maintained a straight line, the airspeed must vary during a descent. However, gliders can make a nearly perfect steady and straight descent due to their very low sink rate.

3.6 Vertical Maneuvers

Whenever the lift is different from that required to balance the gravity component normal to the flight direction, a curved trajectory is immediately produced. Such a curved flight in the vertical plane is termed a *vertical maneuver*. The forces in a vertical maneuver are unbalanced and hence an equilibrium does not prevail, as shown in Fig. 3.6. Therefore, the velocity vector keeps on changing, both in the magnitude and the direction. The changing inclination of the velocity vector is given by the *flightpath angle*, defined as the angle made by the velocity vector with the horizon (also that made by the lift vector with the vertical), and is shown in Fig. 3.6. The centripetal acceleration for the curved flight is provided by the difference between the lift and the normal component of the weight, whereas the thrust opposes the sum of the drag and the tangential component of weight. Both the components

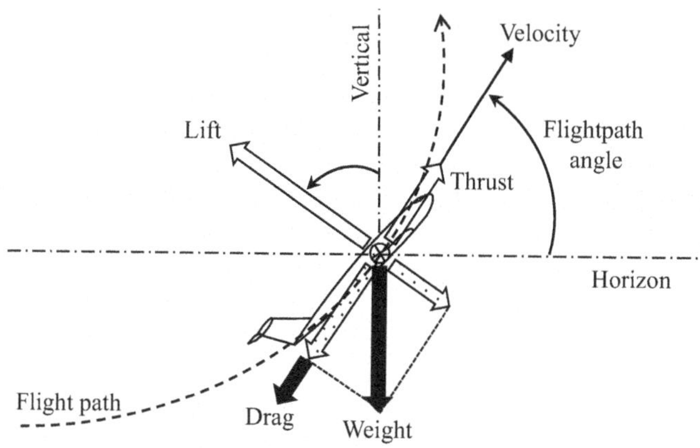

Fig. 3.6 An airplane undergoing a vertical pull-up maneuver, beginning from a level flight condition. The flightpath angle made by the velocity vector with the horizon increases continuously in such a maneuver, and the airspeed correspondingly decreases. The lift must exceed the normal weight component for the required centripetal acceleration. Since the forces are not balanced, an equilibrium does not exist

of the weight keep on changing as the inclination of the flight path is varied through the maneuver. The type of the vertical maneuver produced depends upon the lift-to-weight ratio, called the *load factor*, and is denoted by n. For example, level flight is indicated by $n = 1$. The normal acceleration experienced in a maneuver of load factor n is therefore $n - 1$ times the acceleration due to gravity, g. Hence, a vertical maneuver is commonly indicated by a "g-value," such as a "$2g$" or a "$5g$" maneuver, which means that the normal acceleration is either twice or five times, respectively, the acceleration due to gravity. These numbers would correspond to load factors of 3 and 6, respectively.

The lift must exceed the weight in a vertical *pull-up* maneuver executed from the level flight condition, whose example is depicted in Fig. 3.6. The load factor, n, must exceed unity in such a pull-up maneuver. The normal acceleration experienced during the maneuver is directly proportional to the load factor. As the maneuver progresses from the initial level condition, the airspeed must continuously decline, unless the engine power is increased significantly to balance the increased tangential weight component. If the engine power is unchanged and a constant load factor is maintained, the airspeed may decrease to the stall speed at some point, and a stall will occur. If the maneuver is initiated at a sufficiently large speed, the stall can be avoided and the vehicle can reach a fully inverted position at the maximum altitude point, with a speed higher than the stall speed. The point at which the flightpath angle becomes 180° (implying an inverted flight condition) is called the *top of the loop*, because a complete loop can be now executed with the vehicle finally exiting in a level condition (i.e., zero flightpath angle). Flying a loop is a popular aerobatic maneuver in the airshows. However, since the energy is being continuously lost during a loop, the altitude at its exit is always lower than that at the entry. Many accidents have taken place in performing a vertical loop where the pilots did not make the allowance for the net drop of altitude.

Although it is theoretically possible to maintain a constant energy during a vertical maneuver in which there is no net altitude change, it would require a continuous adjustment of both the engine power and the angle-of-attack through the maneuver, which is difficult to achieve in practice. Such a constant energy maneuver is said to be *conservative*, and requires that the thrust must balance the drag at all the points of the flight path. Since the airspeed as well as the flightpath angle rapidly changes with the time in a vertical maneuver, so does the drag, and the pilot would have to struggle constantly with the throttle levers and the pitch control (to be discussed later) such that the thrust is always kept exactly equal to the drag.

Another theoretical maneuver that some textbook authors like to analyze is that in which a constant airspeed is always maintained. This is even more impractical than a conservative maneuver, because it requires the thrust should exceed the weight and even be reversed in direction at some points on the trajectory. A careful analysis of a constant-speed maneuver reveals that when the flightpath angle reaches 90°, the thrust has to balance the weight plus the drag of the aircraft. On the opposite side of the loop, the vehicle is traveling vertically downward (flightpath angle of 270°), and the only way the thrust can balance the weight (minus the drag) is by reversing its direction! Needless to say, both of these requirements cannot be achieved in practice, even by the most powerful fighter type aircraft.

Rather than maintaining a constant airspeed or energy, it is more practical to keep the airplane or glider in the *trim* condition defined by a zero pitching moment. In such a case, the pilot maintains a constant pitch rate (thus a constant normal acceleration and load factor) while maneuvering. Since the airspeed will be always varying in such a flight (decreasing as it climbs, and increasing as it descends) the pilot must be careful to avoid a stall by keeping the pitch rate (thus the load factor) below a certain value.

Vertical maneuverability of an aircraft is indicated by the largest possible load factor it can achieve. Because the lifting load on the wings is directly proportional to the load factor, n, there is a possibility of structural failure when the load factor is increased beyond a limit. Furthermore, the vehicle's occupants can experience a "black-out" condition in a large, positive load factor maneuver. This happens when the heart cannot keep up with the increased demand of pushing the blood upward to the brain at the vehicle's high centripetal acceleration. Therefore, the brain is starved of the oxygen and the person falls unconscious. Wearing a pressurized suit can alleviate this physiological condition by assisting the heart in circulating the blood. Similar physiological and structural limits exist for the negative load factor, where the vehicle is being accelerated downward (i.e., the lift is negative, and the flightpath angle is continuously decreasing). Now the wing structure can be stressed downward (which is not its normal loading condition), and the blood circulation to the brain can become excessive, causing what is called a "red-out" condition.

Most transport category airplanes (such as the subsonic airliners) have a positive load factor limit of 3.5 and a negative limit of -2 for airworthiness certification. In contrast, the fighter type airplanes can go up to a positive load factor of about 11 and a negative load factor of -5 due to their much sturdier structure and pressurized-suited pilots. This means that a fighter pilot can feel his/her weight has increased up to ten times in a vertical pull-up maneuver (remember, $n = 1$ is the level flight condition; therefore $n = 11$ indicates a normal acceleration of ten times that due to gravity).

3.7 Take-Off and Landing

Take-off and landing are the most crucial points in the mission of any airplane and glider as they involve flying in the vicinity of the ground. More than 99 % of all airplane accidents happen while either taking-off or landing. To understand why this is so, we must explain what exactly takes place at these two flight conditions.

Taking off involves accelerating to a speed sufficiently higher than the stall speed while moving in a straight line along the runway. The standard take-off speed is taken to be 20 % higher than the level flight stall speed with the flaps set in the take-off configuration (see Chap. 2). Airplanes take-off under their own engine power, whereas the gliders are either winched forward, or towed by an airplane. The total distance required to take off—called the *take-off distance*—is defined as the length of the runway necessary to accelerate from rest to a velocity that is large enough

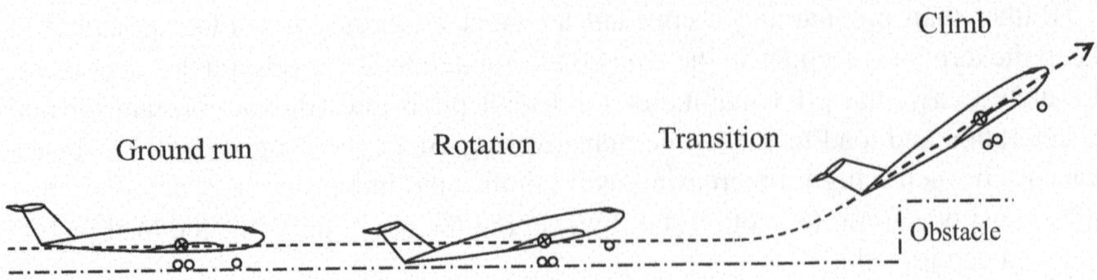

Fig. 3.7 The standard take-off sequence. The airplane accelerates from rest to the take-off speed during the ground run, rotates to increase its angle-of-attack, transitions to a steady climb atop the imaginary obstacle by performing a pull-up maneuver

to establish a steady climb over an imaginary obstacle of a standard height. The obstacle height is conventionally taken to be 50 ft. for single-engine airplanes, and 35 ft. for multi-engine airplanes with the failure of one engine. The one-engine failed take-off (termed *one engine inoperative (OEI)*) is necessary for every multi-engine airplane in order to make a safe landing after suffering an engine failure. Since the take-off is performed nearly at the full engine power, there is an increased possibility of engine failure during take-off. A standard minimum rate-of-climb above the obstacle with OEI must be demonstrated for every multi-engine airplane for airworthiness certification purposes.

Figure 3.7 shows the sequence employed in a standard take-off. Beginning from rest, the airplane accelerates under full engine power during which all its wheels are on the ground. This portion of the take-off is termed the *ground run*, and is continued until the airplane has reached the take-off speed, which is 20 % higher than the level flight stall speed at a take-off flap setting. Since the airplane's angle-of-attach during the ground run is small, it does not yet lift off from the runway. In order to increase the angle-of-attack, the airplane is pitched up after the ground run such that it achieves nearly 80 % of the maximum lift coefficient. The process of pitching the nose upward is called *rotation*, and takes about 2–3 s for a large airplane due to its moment of inertia. During the rotation phase, the speed is nearly constant (the take-off speed), while the nose has lifted from the runway. Having achieved the necessary angle-of-attack at the take-off speed, the airplane leaves the runway at the end of the rotation phase. However, since its speed is 20 % higher than the stall speed, and the lift coefficient about 80 % of the maximum (stall) value, the lift at this point is about 15 % higher than the weight. This means that the load factor, n, is approximately 1.15, which immediately sends the airplane into a vertical pull-up maneuver. This phase of take-off, called the *transition*, is crucial for transitioning from a horizontal to an inclined trajectory, which is necessary for establishing a climb on top of the obstacle. The curvature of the flight path in the transition phase is large, but a flightpath angle of less than 10° is necessary to clear the obstacle. This means that the airspeed does not change very much during the transition phase, at the end of which the airplane has reached the correct angle-of-attack for a steady climb. The take-off distance is the total horizontal distance traveled along the runway until the obstacle is cleared.

The largest component of the take-off distance is that required for the ground run. This is the reason why the take-off is always made at full throttle. Some large jet airplanes produce about 5–10 % higher thrust at take-off than what is allowable in extended operation. The take-off flap setting should not be chosen to be too large because it increases the drag significantly, thereby extending the ground run. However, a small to moderate flap setting is necessary to reduce the stall speed, and hence the take-off speed.

Take-off distance crucially depends upon the weight, the atmospheric density, and the wind direction. The take-off distance roughly increases with the square of the stall speed. As the weight is increased, the stall speed is also increased, thereby requiring a much larger take-off distance if everything else remains the same. When the prevailing atmospheric density at an airport becomes smaller, there is an attendant increase in the stall speed, which causes an increase in the take-off distance. That is why most take-off certification requirements are specified at an increased temperature, which results in a smaller density. Another factor which causes a similar reduction in the density (thus an increase in the take-off distance) is the altitude at which the airport is situated. A mountainous airport will require a larger take-off distance that the one located near the sea level, because it has a lower atmospheric pressure (hence smaller density).

For the shortest distance required, the take-off must be made opposite to the prevailing wind direction. This is called *taking-off into the wind*. The distance traveled along the ground depends upon the ground speed, but the stalling speed is a function of the relative airspeed. Hence, the smallest ground speed for a given airspeed is obtained by flying into the wind.

Landing is the reverse operation of take-off because it involves bring the airplane to rest from a steady descent. The slope of final landing approach is quite shallow (only about 3–5°) for safety considerations, which slightly lengthens the total landing distance. The shallow descent on top of the obstacle is carried out at nearly 30 % higher speed than the stall speed with full flaps. The descending flight is converted into a flight nearly parallel to the runway in a phase called the *flare*, during which the airplane is steadily decelerating as well as pitching upward. At the end of the flare, the airplane has slowed down to nearly 10 % larger velocity than the stall speed. This speed is called the *touchdown speed*. The main wheels of the airplane first come into contact with the runway immediately after the flare at a large angle-of-attack corresponding to the touchdown speed. This is followed by a *free roll* in which the airplane is traveling forward with the main wheels on the ground, and rotating such that the nose is continuously being lowered. During the free roll, the speed of the airplane is allowed to dissipate by ground friction and drag, but the brakes are not applied. As soon as the nose wheel comes into contact with the ground at the end of the free roll, the brakes and retardation devices are applied to bring the airplane quickly to rest. This last phase of landing is called the *braking phase*.

The retardation devices deployed during the braking phase include spoilers and airbrakes. These are flat-plate like devices that extend nearly normal to the flow, thereby increasing the drag enormously. The spoilers are placed on the top of the wings to also decrease the lift substantially, such that more of the weight of the airplane is exerted on the ground. Doing so increases the effectiveness of the wheel brakes. Other commonly used retardation devices are the thrust reversers for jet engines, and reversible pitch mechanisms for propellers. About 15–20 % of the engine's thrust can be applied in the reverse direction in either of these ways, and acts like a brake to shorten the overall landing distance.

When an engine failure takes place during a take-off, the pilot has the option to either continue the take-off, or to apply the brakes and the retardation devices to stop the airplane in the remaining runway. Which of the options is taken depends only upon the speed at which engine failure occurs. There is a critical speed called the V_1 speed, at which the failure of one engine produces the same required distance for both the options. If the actual engine failure occurs at a speed greater than the V_1 speed, the take-off is continued and the airplane goes around and comes back for a normal landing. However, if the speed at which the failure occurs is smaller than V_1, then the brakes are applied immediately, bringing the airplane to rest. It is thus crucial for a safe take-off to calculate the V_1 speed in advance, for a given take-off weight and the prevailing atmospheric density. A co-pilot usually watches the airspeed indicator and calls out "Vee-one" as soon as the critical speed is crossed during the take-off.

3.8 Degrees-of-Freedom and Control Surfaces

For most flight purposes, airplanes and gliders are assumed to be rigid bodies. As we saw in Chap. 1, a rigid body has six degrees-of-freedom relative to a special point called the center of mass. These degrees-of-freedom are the forward, downward, and sideways translation of the center of mass, and the rotations about three mutually perpendicular axes passing through the center of mass. Since all airplanes and gliders have a plane of symmetry, the three axes are defined with respect to that plane, and constitute a reference frame with the origin at the center of mass, as shown in Fig. 3.8.

The rotation in pitch is defined to take place about an axis perpendicular to the plane of symmetry, and passing through the center of mass. This axis is called the *pitch axis*, and is shown in Fig. 3.8 as the line *OB*, with *O* being the center of mass. The change in the orientation of the vehicle in such a rotation is termed the pitch angle, and its rate of change is called the pitch rate. The moment applied about the airplane or glider about the pitch axis is called the pitching moment. The directions "forward" and "downward" are defined along two mutually perpendicular axes normal to the pitch axis, and lying in the plane of symmetry (Fig. 3.8) and are termed the *roll axis* (also the *longitudinal axis*), *OA*, and the *yaw axis*, *OC*, respectively. A rotation of the aircraft about the roll axis is termed roll, and that about the yaw axis is termed yaw, with the corresponding rates and moments.

In order to control the motion of the airplanes and gliders, aerodynamic forces should be applied to them at will. This is carried out by control surfaces. A *rudder* is

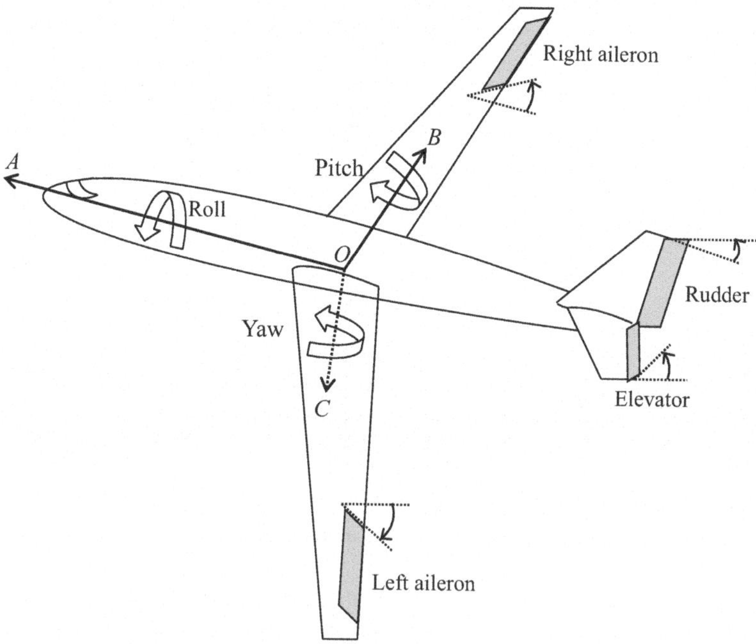

Fig. 3.8 Longitudinal dynamics and the associated frame of reference, *OAC*, in the plane of symmetry

the control surface on the vertical tail that is used for controlling the flight direction by applying a yawing moment on the vehicle. A left (or right) deflection of the rudder creates a yawing moment to the left (or right), moving the nose of the aircraft in the desired direction. An *elevator* is a similar control surface on the horizontal tail or canard, which creates a pitching moment, moving the nose either up, or down as required. A pair of control surfaces mounted on the wing and deflected in opposite directions are called the *ailerons*. The ailerons generate a rolling moment, by which the vehicle can bank either left or right, in order to make a turn in that direction. Figure 3.8 shows the control surfaces of a fixed-wing aircraft (i.e., airplane or glider). Apart from the aerodynamic control surfaces, the thrust of an airplane's engines can be used as an additional control input. The thrust is controlled by the throttle.

The coupled translational and rotational motion of the airplane that lies entirely in the plane of symmetry is termed *longitudinal dynamics*, and comprises the pitching rotation about the center of mass, and the forward and downward translations of the center of mass. The remaining degrees-of-freedom (sideways translation, roll, and yaw) constitute the *lateral–directional dynamics*. Because a pure longitudinal motion does not generate the aerodynamic forces that can excite the

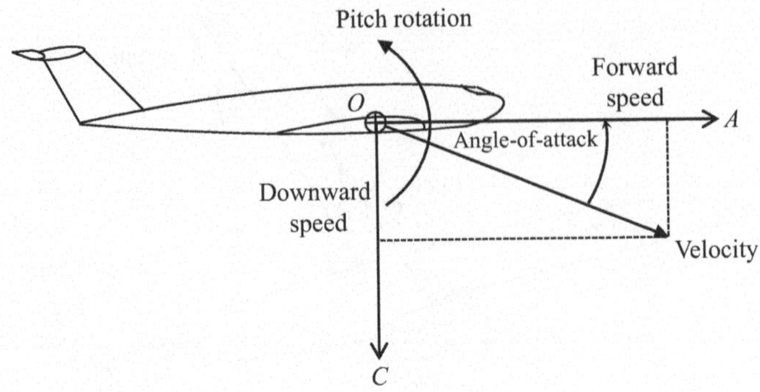

Fig. 3.9 Longitudinal dynamics and the associated frame of reference, *OAC*, in the plane of symmetry. The third axis, *OB*, is the pitch axis perpendicular to the plane of the figure

lateral–directional degrees-of-freedom, the stability and control of an airplane (or glider) in longitudinal dynamics can be analyzed separately from the lateral–directional motion.

Figure 3.9 depicts the variables of longitudinal motion. The aircraft's velocity resolved in the plane of symmetry in the frame *OAC* gives rise to a forward speed along *OA*, and a downward speed along *OC*, as shown in Fig. 3.9. Here, the pitch axis, *OB* is normal to the plane of the figure. The angle-of-attack of the aircraft is defined as the angle made by the velocity vector with the lateral plane, *OAB*. If the longitudinal axis is parallel with the wing's average chord, then this angle would be the same as the angle-of-attack of the wing. However, the wing could have a small incidence with respect to the longitudinal axis. In such a case the two angles-of-attack would differ from each other by an amount equal to the wing's incidence. As discussed next, the elevator and throttle inputs provide a control of an airplane's longitudinal dynamics.

3.9 Speed and Altitude Control

The power required for a steady and level flight depends upon the weight, the atmospheric density, the wing's planform area, and the aerodynamic shape of the airplane. For a given airplane, only the weight and the density govern the variation of the required power. With a weight determined by the payload, fuel, and the crew, the required power varies only with the atmospheric density, which is determined by the altitude. It is thus possible to fly at various altitudes by changing the engine power—which determines the thrust—and by varying the angle-of-attack (which determines the lift and drag), until an equilibrium of the various forces is achieved. While maintaining an equilibrium in the presence of small atmospheric disturbances is performed by small angle-of-attack variations alone, the task of changing the equilibrium condition to a new pair of airspeed and altitude necessary for steady and level flight is more demanding, and requires both the thrust and the angle-of-attack to be adjusted simultaneously. Therefore, every airplane must have independent mechanisms for changing the thrust and the angle-of-attack.

The angle-of-attack is controlled by applying an aerodynamic pitching moment about the airplane's center of mass. This pitching moment is produced by moving the elevator, which is usually mounted on the horizontal tail. The elevator can be either the trailing part of the tail hinged about a fixed axis, or the entire tail itself can be moved as an elevator (the *all-moving* or the *flying* tail). Figure 3.10 shows a schematic diagram of how the pitching moment is applied by elevator deflection. A pitch control lever in the cockpit, called the *stick*, is linked to the elevator—either directly through cables and pulleys, or via an electrohydraulic actuator—such that its fore and aft movement causes the elevator to deflect either up or down. The elevator deflection is the angle between the elevator chord and the chord of the horizontal tail at a given spanwise location. By convention, a pulling of the stick by the pilot causes a nose-down deflection of the elevator (Fig. 3.10). This creates a decreased lift on the tail, thereby applying a nose-up pitching moment on the airplane about its center of mass. Similarly, a pushing of the stick generates a nose-down pitching moment. Since the tail is located at a significant distance from the center of mass, even a small change in the tail's lift can create an appreciable pitching moment due to the essentially large moment arm. Both the sides of the elevator must have the same deflection in order to maintain symmetry.

Some airplane's have a *canard* instead of a tail, which is a tail-like surface located forward of the wing. The elevator is then either a part of the canard, or the entire canard. There are some high-speed aircraft without a tail or a canard, e.g., the French *Mirage* fighters and the erstwhile Anglo–French *Concorde* supersonic airliner. In such a tail-less design, the pitching moment is provided by control surfaces called *elevons* located at the trailing edge of the wing. If the left and right elevons are deflected symmetrically, they cause a pitching moment to be generated by the wing. Of course, for a better effectiveness, the moment arm of the elevons should be reasonably large. This is ensured by a small aspect ratio and a high leading-edge sweep angle of the wings in a tail-less airplane. If the left and right elevons are deflected in opposite directions, they create a rolling moment about the center of mass like a pair of ailerons (to be discussed later).

In a level flight, the changes in the airspeed are produced by the thrust and the drag, while those in the altitude are generated by the lift. Since the drag (and thrust) are about one-tenth of the lift for the steady and level flight of a well-designed airplane, a variation in the airspeed takes a much longer time to produce than that in the altitude. Consequently, a pilot relies mainly on the pitch control which determines the angle-of-attack (thus the lift) in order to control the airspeed required for level flight at a given altitude.

The engine power controls the thrust, and is determined by the throttle input supplied by the pilot. If there are several engines, each of them must be producing exactly the same thrust in order that there is no lateral torque on the airplane. Otherwise, the airplane will depart from a steady and level flight. That is why each

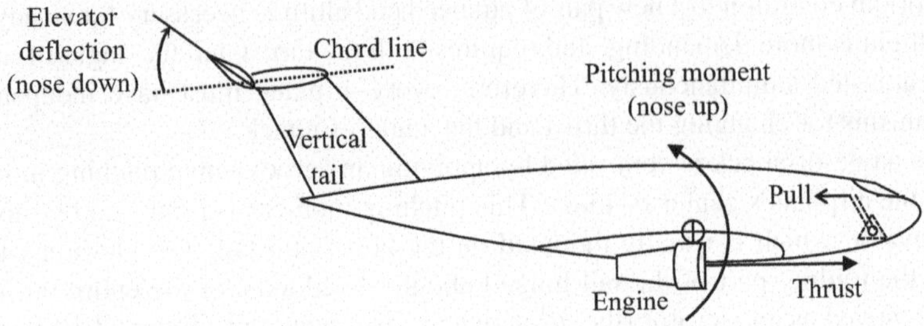

(a) Side view.

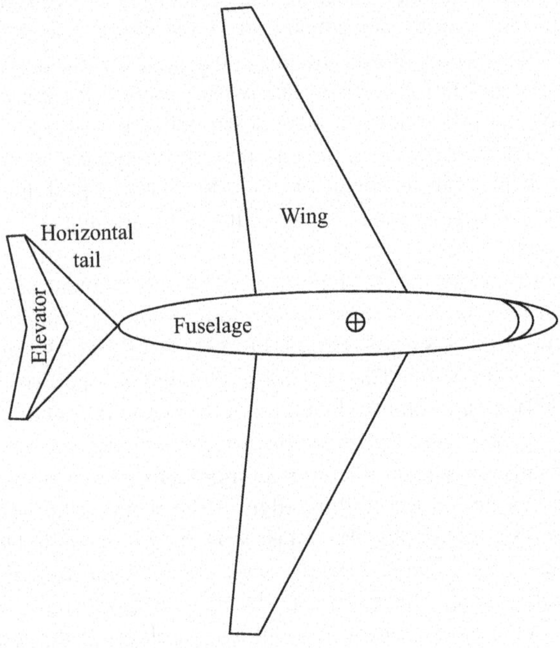

(b) Top view.

Fig. 3.10 Schematic diagram showing the pitch control of an airplane with an elevator. (**a**) *Side view* depicts the situation when a pulling on the stick creates a nose-up pitching moment on the airplane due to a decreased lift on the tail. Engine thrust is separately controlled by a throttle quadrant (not shown). (**b**) *Top view* indicates the relative locations of the wing, the horizontal tail, and the elevator, showing the large moment arm from the tail to the center of mass (+)

engine must be throttled separately. A four-engined airplane such as the Boeing-747 has four throttle levers to be adjusted every time the thrust has to be varied. This is not only cumbersome for the pilot, but also takes some time to implement, mainly due to the engine dynamics involved. A large jet engine, such as that of a Boeing-747, could take several seconds to produce the variation in the thrust demanded by the throttle input. Even if the thrust could be changed instantly, it would still take a long time for the airspeed to change to a desired new value. Therefore, controlling the airspeed is rarely performed by the throttle, and instead the pitch control is utilized. However, the vertical speed (or the rate of climb/ descent) is much more easily controlled by the throttle rather than the pitch control, because it is directly determined by the "balance of power" between thrust and drag. Thus, a quick change in the altitude (at nearly the same airspeed) is performed by making throttle adjustments.

While discussed here in the context of steady and level flight, the pitch and throttle inputs can be applied to control any flight situation. However, due to their symmetric nature, these inputs can control only the motion of the airplane in the plane of symmetry.

3.10 Static Longitudinal Stability

When an airplane is traveling in a fixed direction at a constant airspeed and altitude, its aerodynamic forces are balanced by the weight and the thrust. If an atmospheric gust—or an inadvertent pilot input—causes a slight disturbance from the steady and level flight condition, the airplane must have a natural property to return to the equilibrium. The property of regaining an equilibrium once disturbed from it is termed *stability*. Since stability must be an inherent characteristic of an airplane, it should not depend upon the actions of the pilot. In other words, pilot inputs should not be required to maintain the airplane in equilibrium.

Stability has two essential features: (a) static stability and (b) dynamic stability. If a spring-like restoring force (or moment) can be naturally generated by an airplane once it is disturbed from the equilibrium, then the airplane is said to be statically stable, because its motion always tends toward the equilibrium. Of course, as in the case of the spring, there is no guarantee that an airplane possessing static stability will actually settle down to the equilibrium condition. Instead, it might display an oscillating behavior about the equilibrium. Therefore, merely having static stability is not sufficient for an overall stable behavior. The airplane must also have a dynamic stability by which it settles down to the original equilibrium flight condition, after being initially disturbed. If an oscillatory response to any initial disturbance persists for a long time, then the airplane is said to have poor dynamic stability. Hence, dynamic stability is the property by which any oscillations are quickly damped out to zero.

If the static stability is absent, there is not even a tendency to move toward the original equilibrium. Consequently, an ever increasing departure takes place from the equilibrium for a statically unstable airplane, and the question of dynamic stability does not arise since no oscillations are involved.

We will focus on longitudinal stability in this section, which is defined as the property to return to the equilibrium condition defined by a pair of airspeed and altitude in a steady and level flight, after being disturbed by a gust (or pilot input) in the plane of symmetry. However, before discussing stability, it is important to consider the airplane as an extended body (rather than as a point) on which the different forces are applied at different points. Since the lift, drag, weight, and thrust, are acting at different points on the airplane, they are also creating moments about the center of mass. For equilibrium condition of steady and level flight, not only the forces must be balanced, but they should also produce a zero net pitching moment about the center of mass. If this were not so, the airplane will begin rotating about the pitch axis.

The aerodynamic shape of the airplane is such that it has an unbalanced nose-down pitching moment when the lift is zero. This is due to the cambered shape of the wing airfoil, which produces a non-zero lift when the angle-of-attack of the wings is zero. Consequently, at a positive angle-of-attack measured from the wing's mean chord plane (see Fig. 3.11), the lift is much higher than what it would be for an uncambered (symmetrical) airfoil. For achieving the moment balance in such a case, a tail (or a canard) is necessary. In order to understand this, let us consider a statically stable airplane with respect to angle-of-attack disturbance. The locations of the points at which the lift of the wing and the tail are acting relative to the center of mass now become crucial in this analysis.

We saw in Chap. 2 that the *aerodynamic center* is a special point on the wing about which there is no variation in the pitching moment, when there is a change in the angle-of-attack. If we take the wing's lift to be a force applied at the aerodynamic center, then we must also add a nose-down pitching moment about that point which is constant with respect to the angle-of-attack. This is the same pitching moment

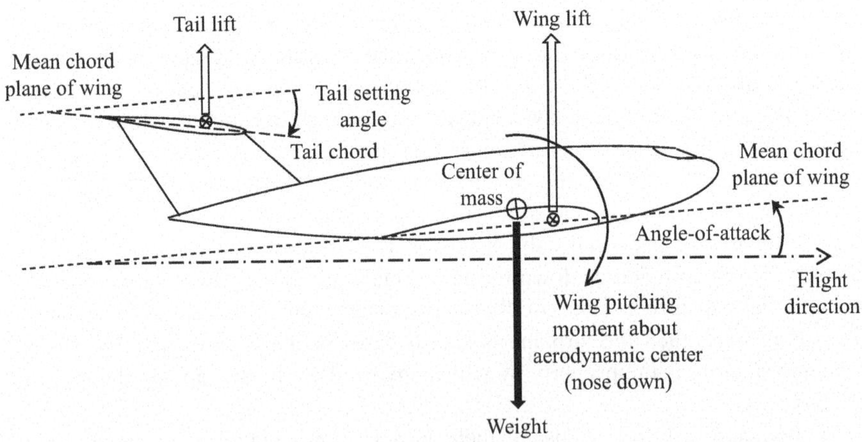

Fig. 3.11 Force and moment equilibrium (trim) at a steady and level flight condition with a positive angle-of-attack of the wing. The wing and the horizontal tail produce lift about their respective aerodynamic centers, whose sum must balance the weight. Due to its cambered shape, the wing also produces a nose-down pitching moment about its aerodynamic center, which remains constant with respect to the angle-of-attack. Note that the tail is at a negative incidence (tail setting angle) with respect to the wing, which is exaggerated here for clarity

which is produced by the wing when its angle-of-attack is zero. Similarly, the tail has its own aerodynamic center. However, because the tail acts as the mount for the elevator, it must produce equal magnitudes of both negative and positive lift for control purposes (see above). Therefore, the tail usually has a symmetrical airfoil, and hence its pitching moment about the aerodynamic center is zero. Figure 3.11 shows the wing and tail lifts acting at their respective aerodynamic centers, along with the constant nose-down pitching moment about the wing's aerodynamic center. The situation of initially balanced forces, as well as the zero net pitching moment about the center of mass due to those forces is also depicted in Fig. 3.11. Such a force and moment balance is called the *trim* condition. Note that only the wing's and the tail's lift and the wing's constant nose-down pitching moment about the aerodynamic center contribute to the net pitching moment about the center of mass. The weight has no moment contribution, as it passes through the center of mass. For simplicity, we have neglected the small pitching moment caused by the engine thrust, because it acts very near the center of mass in a well-designed airplane.

If the airplane is statically stable with respect to an angle-of-attack disturbance, it must produce a nose-down pitching moment whenever the angle-of-attack is increased. Furthermore, the magnitude of the corrective nose-down pitching moment must be proportional to the change in the angle-of-attack. A larger variation in the angle-of-attack would then generate a larger restoring moment, and vice versa. Such a pitching moment would tend to restore the airplane toward the original (smaller) angle-of-attack at which an equilibrium had prevailed. By having the magnitude of the pitching moment proportional to the change in the angle-of-attack will ensure that the corrective action is applied until the equilibrium is restored, i.e., when both the angle-of-attack disturbance and the net pitching moment are zero. Taking the nose-up pitching moment to be positive, we thus require that the slope (or the rate of change) of the pitching moment with respect to the angle-of-attack must be negative for a static longitudinal stability. This idea is depicted in Fig. 3.12. The lines *AB* and *CD* denote statically stable aircraft.

If the slope of the pitching moment vs. angle-of-attack straight line were to be positive (the line *BE* in Fig. 3.12), then the airplane would tend to move toward a larger angle-of-attack after every increase in the latter, which in turn, will cause an even larger nose-up pitching moment. Clearly, this would be a statically unstable situation.

The direct proportionality of the restoring pitching moment with the change in the angle-of-attack arises naturally, because both the wing's and the tail's lift are linear functions of the angle-of-attack. For simplicity, the discussion given here is very basic, and does not include the aerodynamic effects of the fuselage, and the wing's downwash on the tail (or that of the canard on the wing). Furthermore, it is assumed that the angle-of-attack remains small, for which the lift and the pitching moment vary linearly (i.e., as a straight line) with the angle-of-attack. For a large change of the angle-of-attack, the aerodynamic forces and moments become nonlinear functions of the angle-of-attack, and the present analysis would not be valid.

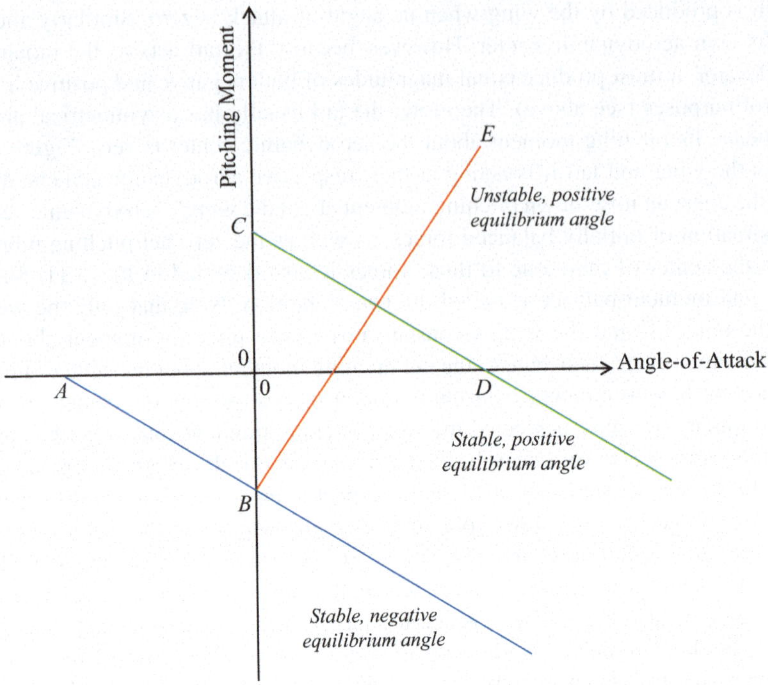

Fig. 3.12 Pitching moment vs. the angle-of-attack variation. All airplanes and gliders have positively cambered wings, which produce a negative (nose-down) pitching moment at zero angle-of-attack. In order to be statically stable, as well as to fly at a positive angle-of-attack, a tail (or canard) is required, which provides a positive pitching moment when the wing's angle-of-attack is zero. The symmetrical tail airfoil must be set at a negative incidence with respect to the mean chord of the wing for this purpose (see Fig. 3.11)

Because the straight line plotted for the restoring pitching moment vs. angle-of-attack variation must have a negative slope for static stability, the airplane must have a positive value of the pitching moment when the angle-of-attack is zero. This is necessary if an equilibrium (that is, zero pitching moment) is to be achieved at a positive angle-of-attack, as shown in Fig. 3.12 by the line *CD*. However, since the cambered wing creates a negative (nose-down) pitching moment at a zero angle-of-attack, this is not achievable by the wing alone. The line *AB* in Fig. 3.12 indicates precisely this situation, where the point *B* is at a negative pitching moment, and the point *A* denotes a possible equilibrium at a negative angle-of-attack. What is required is a tail that can apply a negative lift (i.e., positive pitching moment) when the wing is operating at a zero angle-of-attack. Conversely, if a canard is used rather than the tail, it must create a positive lift in the same situation. Figure 3.12 shows how a tail (or canard) can cause the net pitching moment about the center of mass to be come positive (i.e., change the straight-line's intercept point from *B* to *C*) when the wing is at a zero angle-of-attack. This is achieved in practice by rigging the tail such that it is at a negative (nose-down) angle with respect to the wing, as shown in Fig. 3.11. This small negative angle (usually 2–3°) is called the *tail setting angle*. Since the tail has a symmetric airfoil, its lift will be always negative when the wing's angle-of-attack is zero. In contrast, a canard must be set at a small positive angle relative to the wing.

The *margin* of longitudinal static stability is indicated by the magnitude of the (negative) slope of the pitching moment vs. angle-of-attack line. When the slope is more negative, the aircraft is more stable. Since the location of the center of mass is the only variable which determines the magnitude of the pitching moment for a given angle-of-attack, it governs the static stability. There is a particular location of the center of mass on each airplane for which the airplane is neutrally stable, i.e., it has a zero slope of the pitching moment variation with the angle-of-attack, which implies a zero static stability. This position of the center of mass is called the *neutral point*. If the airplane is loaded such that its actual center of mass lies aft (more toward the tail) than the neutral point, then such an airplane will be statically *unstable*. A center of mass located forward of the neutral point produces a statically stable configuration. Thus, the distance of the center of mass from the neutral point is a convenient way of measuring the margin of static stability of an airplane (or glider). Since it would be very difficult for a pilot to maintain an unstable aircraft in equilibrium (it would be like balancing a stick on one's palm), all aircraft must possess a positive static stability margin. To put things in perspective, the first powered airplane to fly successfully—the Wright *Flyer*—was a statically unstable design (denoted by the line *BE* on Fig. 3.12), and thus a nightmare to fly. The only advantage a statically unstable design offers is of better maneuverability, which is ability to quickly depart from an equilibrium condition.

Most transport type airplanes must have a minimum static stability margin of 10–15 % of the wing's mean chord. However, many fighter airplanes—like the Lockheed *F-16* and the Sukhoi *Su-30*—are specially designed for higher maneuverability (which is the property opposite to stability), and thus have much smaller (even negative) static stability margins. However, such fighter aircraft have a dedicated automatic stabilization system to keep them in equilibrium, without requiring pilot inputs.

3.11 Longitudinal Control

Since the elevator can apply a large pitching moment, thereby changing the pitch rate, pitch angle, the downward and the forward speed of an airplane or glider, it is the primary longitudinal control device. When flying in an equilibrium condition at a given altitude, the airspeed of flight can be adjusted by changing the angle-of-attack via a pitching rotation of the aircraft carried out by the elevator input. The trim condition of a zero pitching moment thus occurs at a new value of the angle-of-attack (thus the airspeed). The change in the trim angle-of-attack for a statically stable aircraft is shown in Fig. 3.13. Note that the elevator input only changes the intercept of the pitching moment line while maintaining its slope, such that

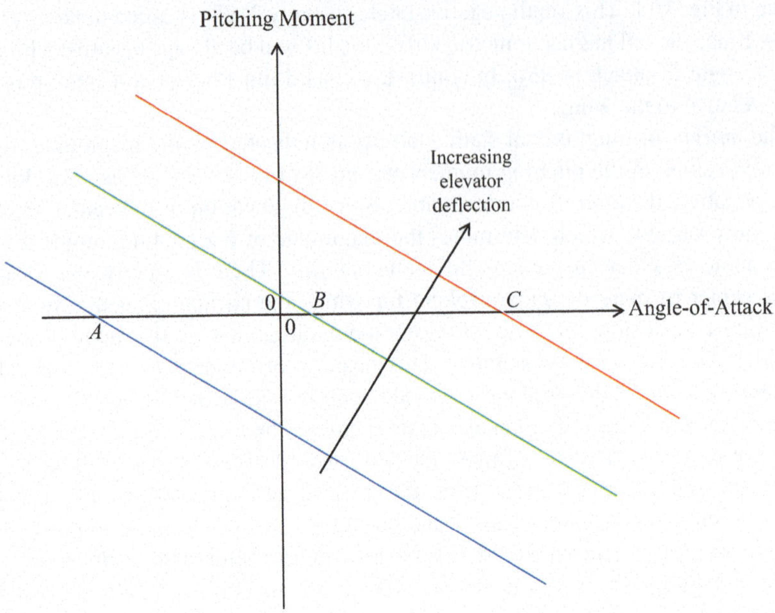

Fig. 3.13 Change in the pitching moment vs. the angle-of-attack line caused by the elevator deflection, in order to fly in a trim condition at various angles-of-attack (*A*, *B*, and *C*). Note that only the intercept of the line is changed and not its slope, which is a constant for a given location of the center of mass

equilibrium condition is achieved at different angles-of-attack, indicated in Fig. 3.13 by points *A*, *B*, and *C*. A similar control can be provided in a maneuvering flight, where the trim condition has to be maintained for keeping the pitch rate a constant.

If an airplane has the possibility of changing the location of its center of mass in flight (by shifting its load), a change in the equilibrium could be provided by changing the slope (rather than the intercept) of the pitching moment vs. angle-of-attack line. However, such an airplane would be very dangerous to fly, because it would be stable at some trim points, and unstable at the others. For this reason, any shift in the center of mass must be avoided in flight.

3.12 Dynamic Longitudinal Stability

All airplanes and gliders are designed to fly in an equilibrium condition of approximately a straight line and a constant airspeed. However, such an equilibrium is not possible if every time a disturbance is encountered, the airplane begins oscillating in pitch, altitude, and airspeed, and continues to do so for a long time. A statically stable aircraft will tend to oscillate about the equilibrium point, because its aerodynamic forces and moments will always move it toward the equilibrium in a spring-like manner. Just as a car on encountering a bump on the road oscillates up and down on its suspension, so does an airplane. But the suspension on which the airplane "rides" is invisible. Continuing the comparison with a car, there should be "shock-absorber" present to quickly damp out the oscillations in the pitch, forward speed, and downward speed; otherwise, a smooth ride would be impossible. If an adequate damping is present in all the degrees-of-freedom, then the longitudinal motion said to be dynamically stable.

The term *damping* refers to any force or a moment which opposes the *rate of change* of a particular degree-of-freedom, and therefore saps the energy of the motion with time. The simplest form of damping is when the opposing force (or moment) is directly proportional to the rate of change in a given degree-of-freedom.

The aerodynamic forces provide a natural damping effect to the aircraft in case of an oscillation in any degree-of-freedom. Suppose there is a sudden increase in the forward velocity due to an atmospheric gust. The increased airspeed will immediately translate into an increased drag on the aircraft, causing its speed to become smaller. Since drag dissipates the energy in forward motion, it also provides a natural damping effect. However, drag is a much smaller force (about ten times so) than the lift in a well-designed aircraft. Therefore, any damping produced by the drag is also much smaller than that possible if the lift were opposing the rate of change of motion.

The small damping produced by the drag force to a change in the forward speed produces a dynamically damped motion called the *phugoid mode*. This is a lightly damped, long-period oscillation in the airspeed, altitude and pitch angle, sending the aircraft into a nearly constant amplitude roller-coaster ride through the air. The angle-of-attack does not vary in such a motion, because the aerodynamic force exciting the motion (drag) is applied tangentially to the flight path. The phugoid mode can be excited by the pilot by varying the throttle without changing the angle-of-attack, and can persist for several minutes, unless damped out by pilot inputs.

In contrast, if the airspeed is not disturbed, but the angle-of-attack is suddenly changed, there would be another kind of longitudinal dynamic oscillation produced. The pitching moment and lift force would govern the new type of oscillation, which are much larger effects than the drag. Consequently, the oscillation in the angle-of-attack is much better damped, and for most aircraft persists for only 1–2 s. This mode of longitudinal oscillation is called the *short-period mode*, and is normally excited by a sudden elevator input. The damping provided by the tail (or canard) plays the most important role in the short-period mode. As the airplane is pitched nose-up during the oscillation, the tail is rotated downward at a proportionally large rate due to its large distance from the center of mass. In doing so, the tail encounters an increased angle-of-attack in proportion to the aircraft's pitch rate. The increased tail lift immediately translates into a nose-down pitching moment, thereby damping the pitching motion. A similar damping moment is produced by a canard (but with the lift in the opposite direction to the tail).

If an aircraft possesses neither a tail nor a canard, its short-period damping would be rather small. In such a case, the wing elevons (see above) can be deflected 180° out of phase with the pitching oscillation by an automatic control system to provide increased damping. A *rate gyro* can sense the pitch rate, and an electrical signal proportional to the sensed pitch rate can be sent via an amplifier to drive the elevon motor in the opposite direction. Such a system is commonly used in the tail-less aircraft, and is termed a *pitch stability augmentation system*. A detailed discussion of automatic control systems is, however, beyond our present scope.

3.13 Stalling Characteristics

Every airplane and glider ceases to fly when the airspeed falls below the stall limit. As explained earlier, a stall takes place when the demand for an increased lift at such a low speed cannot be met by increasing the angle-of-attack. An excessive flow separation then leads to an abrupt loss of lift. If the airspeed cannot be increased quickly, a large and sudden drop of the altitude takes place, which can be dangerous. How the aircraft behaves during the stalled flight is therefore an important part of its flight characteristics.

If there is a natural tendency in the aircraft to drop its nose while stalling, then the angle-of-attack can be decreased below the stall limit, and a recovery can be made without a large altitude drop. The variation of the zero-lift pitching moment with the angle-of-attack at stall (Fig. 2.6b) thus assumes a great significance. A cambered wing, such as the one analyzed in Fig. 2.6b, has a reduction (turning down) in the pitching moment as the angle-of-attack is increased at the point of stall. This is a stable behavior, and causes a negative (nose-down) pitching moment applied to the aircraft as it stalls. Certain airfoils having a sharp leading edge can display the opposite tendency of pitching up as they stall, and are thus said to have an unstable stalling behavior.

However, the tail can dramatically change the stalling characteristics of an aircraft. If the tail stalls at a smaller angle-of-attack than the wing, it creates a nose-up pitching moment at stall, which tends to further increase the wing's angle-of-attack. This is an unstable stall, where the aircraft remains in a stalled condition, and a recovery is not possible because the elevator control is ineffective. Therefore, a tail must be designed such that it has a much larger stalling angle-of-attack than the wing, which would allow the application of the elevator input to decrease the angle-of-attack of the wing after a stall is encountered.

For a better stalling behavior, it is much better to have a canard rather than a tail. This is because the canard produces a positive lift for maintaining the trim condition, and will naturally stall before the wing as the angle-of-attack is increased. The canard stall will apply a nose-down pitching moment to the aircraft, causing a reduction in the angle-of-attack of the wing, thereby preventing a wing stall. A canard configuration is therefore said to be virtually *stall proof*.

Stalling characteristics also include the lateral–directional motion, which is considered in the next chapter.

Chapter 4
Flight of Airplanes and Gliders: Horizontal Plane

4.1 Introduction

Maneuvering in a horizontal plane becomes necessary for airplanes and gliders, whenever there is a need to change the direction of flight. Actually, this can also be done by performing a vertical maneuver. However, as we have seen in the last chapter, a vertical maneuver requires a large increase in the power (thus fuel expenditure) from that necessary for a level flight. Furthermore, a vertical maneuver carried out for a quick change in the flight direction would put an enormous load on the aircraft structure, apart from being highly uncomfortable for the passengers and crew. For a part of such a flight, the airplane would be flying in an inverted position and the occupants would be upside down, spilling their drinks. They will also experience a large normal acceleration, which would appear to increase their weight many fold. A passenger airliner is not certified for inverted flight due to safety reasons. Therefore, vertical maneuvers are normally avoided for fuel economy, structural safety, and passenger comfort. In contrast, a horizontal turn can be carried out much more efficiently, and without putting a large stress on the structure (and the people on board). When properly executed, a level and coordinated turn goes unnoticed by the passengers and the cabin crew, other than tilting the drinks a little.

Vertical maneuvers are similarly avoided in the gliders (except during an aerobatic display), because they involve a large expenditure of the total energy, resulting in a net loss of altitude every time such a maneuver is conducted. A nearly level coordinated turn can produce the same change in the flight direction with a much smaller loss of altitude.

Due to the very nature of a flight maneuver, an equilibrium condition, in which all the forces and moments are naturally balanced, does not exist. Since a horizontal maneuver is not carried out in the plane of symmetry, there are laterally unbalanced forces and moments produced on the airplane, which must be nullified by pilot inputs. If this is not carried out, the airplane would not be turning at a constant rate

A. Tewari, *Basic Flight Mechanics*, DOI 10.1007/978-3-319-30022-1_4

and altitude. Therefore, stability and control of horizontal maneuver is crucial in a safe operation of any airplane or glider. This chapter analyzes the basic principles behind horizontal maneuvers, including their stability and control.

4.2 Level and Coordinated Turn

Whenever a turn has to be made, a centripetal force must be supplied. Unlike the automobiles, which can turn with the aid of the lateral ground friction, an airplane (or glider) must bank in order to make a turn. In theory, an airplane can carry out a wide, side-slipping turn with the help of the sideforce from the fuselage and the vertical tail. This would require flying with a *sideslip angle*, which is the angle made by the fuselage center-line with the flight direction (see Fig. 4.1). Because the fuselage will be inclined with respect to the flight direction, its drag will greatly increase during such a turn. Furthermore, the engine thrust would also act at an

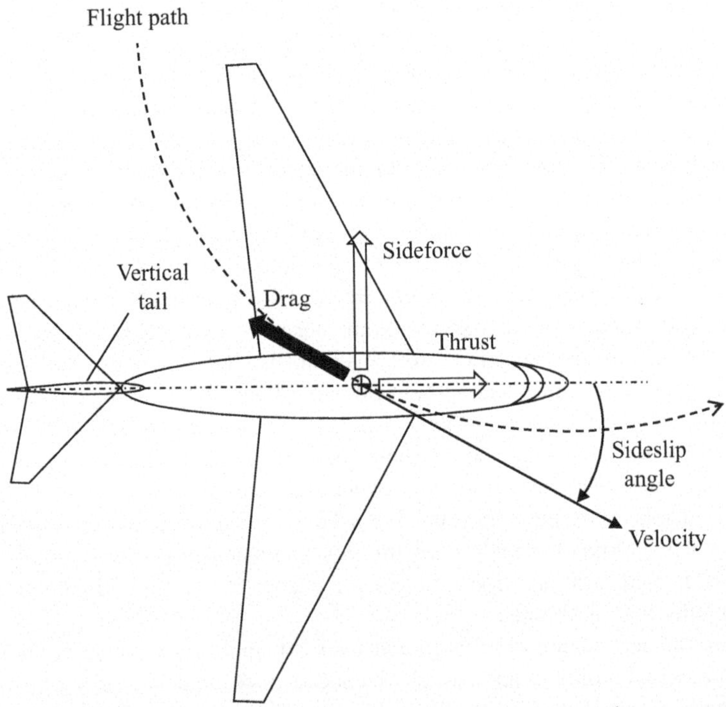

Fig. 4.1 A side-slipping left turn performed by an airplane using a non-zero sideslip angle. The centripetal acceleration required for the turn is supplied by the sideforce generated by the fuselage and the vertical tail. However, by flying at a sideslip angle, the fuselage creates an enormous drag, which cannot be balanced by the thrust. Hence, neither the airspeed nor the altitude would be maintained in such a turn

angle relative to the flight direction. Both of these factors would prevent the thrust balancing the drag in order to maintain a level flight. Therefore, the airplane will rapidly lose altitude while making a turn with the aid of the sideforce. Because the airplane will also slow down appreciably due to the increased drag, a stall can follow, which would be dangerous in an asymmetrical flight situation (to be discussed later). These are primarily the reasons that a side-slipping turn is not carried out in practice.

For the smallest possible drag, an airplane must have its fuselage always aligned with the flight direction while turning. This flight condition of maintaining a zero sideslip angle is called a *coordinated turn*. A coordinated turn also enables the thrust to be applied in the flight direction. Therefore, it is possible to keep the drag exactly balanced by the thrust for maintaining a level altitude while turning. (We recall from Chap. 3 that a level flight at constant speed requires that the power required to overcome drag must be exactly equal to the available thrust power.) There is no sideforce in a coordinated turn, and the centripetal acceleration is supplied by the horizontal component of the lift, which is produced by banking the wings toward the direction of the turn. This situation is depicted in Fig. 4.2, which shows a level, coordinated turn at a constant *bank angle* toward the left. The bank angle is measured between the average plane formed by the wings (i.e., normal to the plane of symmetry) and the horizon, and is the same as the angle between the lift and the vertical direction. As both the lift and the weight are acting normal to the flight path, while the thrust is balancing the drag, there is no net force on the vehicle along the flight direction as well as in the vertical direction; hence its speed and altitude are constant. Since both the altitude and the airspeed are held constant in such a turn, it implies that the total energy is conserved. A level coordinated turn is therefore a conservative maneuver.

A level and coordinated turn must have the lift greater than the weight of the airplane. This can be understood from Fig. 4.2a by considering that the vertical component of the lift must balance the weight in order to maintain an equilibrium of the forces in the vertical direction. Since the vertical component of the lift has to balance the weight, while the lateral lift component must provide the centripetal force for turning, it follows that the total lift must be greater than the weight. Hence, the load factor (see Chap. 3) is always greater than unity for an airplane in a level and coordinated turn.

The increased lift requirement for making the turn translates into a higher angle-of-attack, thus a higher drag. Therefore, engine power must be increased from that required in a straight and level flight. If the pilot fails to increase both the angle-of-attack and the engine power while initiating a turn, the airplane will lose altitude during the turn.

The increased load factor during a level and coordinated turn puts an additional load on the wing structure, and is thus limited for structural and physiological reasons in the same manner as for a pull-up vertical maneuver. The same load factor limits are applicable to the airplane for both horizontal and vertical maneuvers. However, the load factor in a level and coordinated turn rarely cross $n = 2.0$, which corresponds to a bank angle of 60°. Only a fighter type airplane would need to

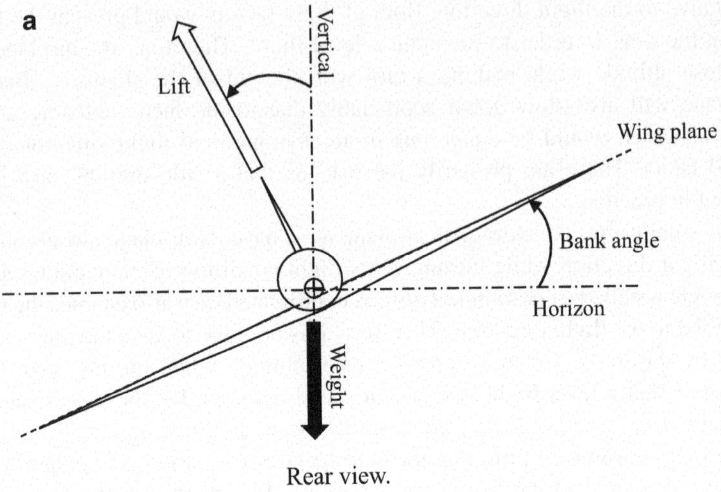

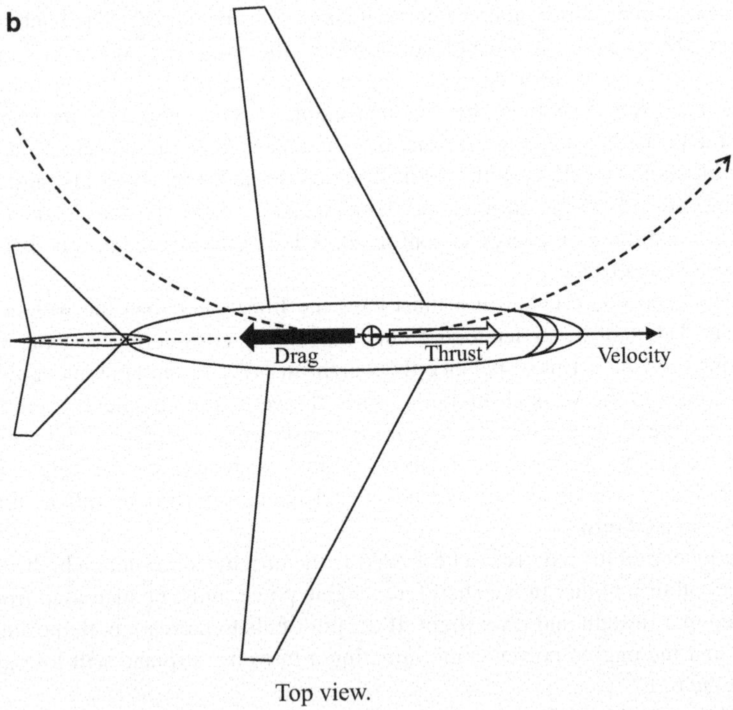

Fig. 4.2 A level and coordinated turn performed by banking the airplane to the left, and then holding the bank angle constant. The sideslip angle is zero. The horizontal lift component provides the centripetal acceleration, while the vertical component balances the weight. The drag is balanced by the thrust, thereby producing a trajectory whose total energy is conserved. The speed and altitude are thus maintained constant during a level and coordinated turn

make a turn steeper than 60°, which might be necessary in an air combat situation. Therefore, horizontal maneuvers are actually less demanding on the structure (as well as the occupants) than vertical maneuvers for a given airspeed and change in the flight direction.

The increased angle-of-attack required for making a coordinated turn from a straight and level flight can take the vehicle closer to the stall limit, if the initial airspeed is already too small. To ensure that a stall does not occur while turning (a dangerous situation!), the airspeed should be above a certain minimum value before executing the turn. Since a correctly executed turn at a constant bank angle has a constant airspeed, a stall can thus be avoided.

4.3 Lateral–Directional Dynamics

Any motion which is not confined to the plane of symmetry can be described by roll, yaw, and sideslip degrees-of-freedom. Since all the three degrees-of-freedom are coupled, we talk of a *lateral–directional* motion in which the three are simultaneously excited. We recall from Chap. 3 that the roll is a rotation of the airplane about the longitudinal (or roll) axis, which is a straight line in the plane of symmetry joining the airplane's nose and tail, and passing through the center of mass. The yaw refers to a rotation of the airplane about an axis which is perpendicular to the longitudinal axis, but lies in the plane of symmetry (and passes through the center of mass). The sideslip is a lateral translation of the airplane's center of mass normal to the plane of symmetry. The displacements in roll, yaw, and sideslip are described by the bank angle, yaw angle, and the sideslip angle, respectively. While the other two angles have been described previously, the yaw angle is the angle made by the longitudinal axis with a horizontal reference direction, and is also termed the heading angle. The lateral dynamics involves the rolling motion, whereas the directional dynamics consists of yaw and sideslip motions. It is easy to understand this terminology, because the direction in which the vehicle's longitudinal axis is pointed is given by the heading (yaw) angle, and the orientation of the longitudinal axis relative to the flight direction is determined by the sideslip angle. A lateral (rolling) motion affects the flight direction, and vice versa.

Associated with the roll, yaw, and sideslip degrees-of-freedom are the aerodynamic rolling moment, yawing moment, and the sideforce, respectively. An airplane or a glider in a sideslip generates all the three of these aerodynamic effects. Similarly, if an airplane is undergoing a rotation in either yaw or roll, it also generates the rolling and yawing moments, in addition to the sideforce. This aerodynamic coupling of the roll, yaw, and sideslip makes it more difficult to design a stability and control system for a lateral–directional motion, than that for the longitudinal case involving pitch rotation and the forward and downward translations. The reason for the increased difficulty is that while the longitudinal motion possesses a natural equilibrium condition in which the pitch rate, and the forward and vertical speeds are held constant, such an equilibrium state does not exist for the lateral–directional dynamics. Consequently, as soon as one of the three angles begins to change, the other two are changed as well. Another way of stating this fact is that the longitudinal dynamics consists of a single rotation and two translations, whereas the lateral–directional motion comprises two rotations and one translation. Even if the two rotational rates were to be constant, the vehicle would be in a non-equilibrium state where all the aerodynamic forces and moments are simultaneously excited.

4.4 Static Lateral–Directional Stability

As in the longitudinal case, an airplane or glider must possess stability in lateral–directional dynamics. The lateral–directional static stability, in which a restoring force (or moment) must be naturally produced upon being disturbed from equilibrium, can be defined in exactly the same manner as static longitudinal stability. Here, the sideslip angle plays the role of the angle-of-attack, and the pitching moment is replaced by rolling and yawing moments. The equilibrium condition is always taken to be a straight-line flight in which the airspeed is constant.

Considering the directional (yawing) motion first, as soon as a sideways velocity disturbance (gust) is encountered, it results in a non-zero sideslip angle, and the vertical tail airfoil sees an increase in its angle-of-attack. This creates a sideforce in the direction of the velocity disturbance as shown in Fig. 4.3a. Due to the sideforce from the vertical tail, a yawing moment is generated on the center of mass in the direction opposite to the velocity disturbance, which tends to move the nose of the aircraft toward the disturbance. In this way, the sideslip angle is decreased. Since the vertical tail has a symmetrical airfoil, the same restoring yawing moment is produced on encountering a disturbance from either the left or the right side. If the tail were absent, the fuselage would tend to yaw in the same direction as the sideways gust (Fig. 4.3a), thereby increasing the sideslip angle. Thus an airplane or glider has a natural *directional static stability* due to the presence of the vertical tail. Such a static stability in yawing motion is also called the weathervane effect, and increases with the size as well as the distance of the tail from the center of mass.

When a sideways gust is applied to an aircraft, it not only experiences a yawing moment, but also a rolling moment. Therefore, a natural static stability in roll is desirable. The production of a rolling moment due to a sideslip is caused by the wing not seeing the same angle-of-attack on its left and right sides. Suppose the sideways gust is encountered from the right side of the aircraft as shown in Fig. 4.3b. In that case, the right wing has a larger angle-of-attack, and thus produces a larger lift compared to the left wing. Why this happens can be explained by the streamlines denoting the side flow component past the aircraft. This flow naturally diverges before reaching the fuselage in order to go around it, and in the process, increases the upward flow component (causing an increase in the angle-of-attack) on the part of the wing it encounters first (in the present example, the right wing). The difference

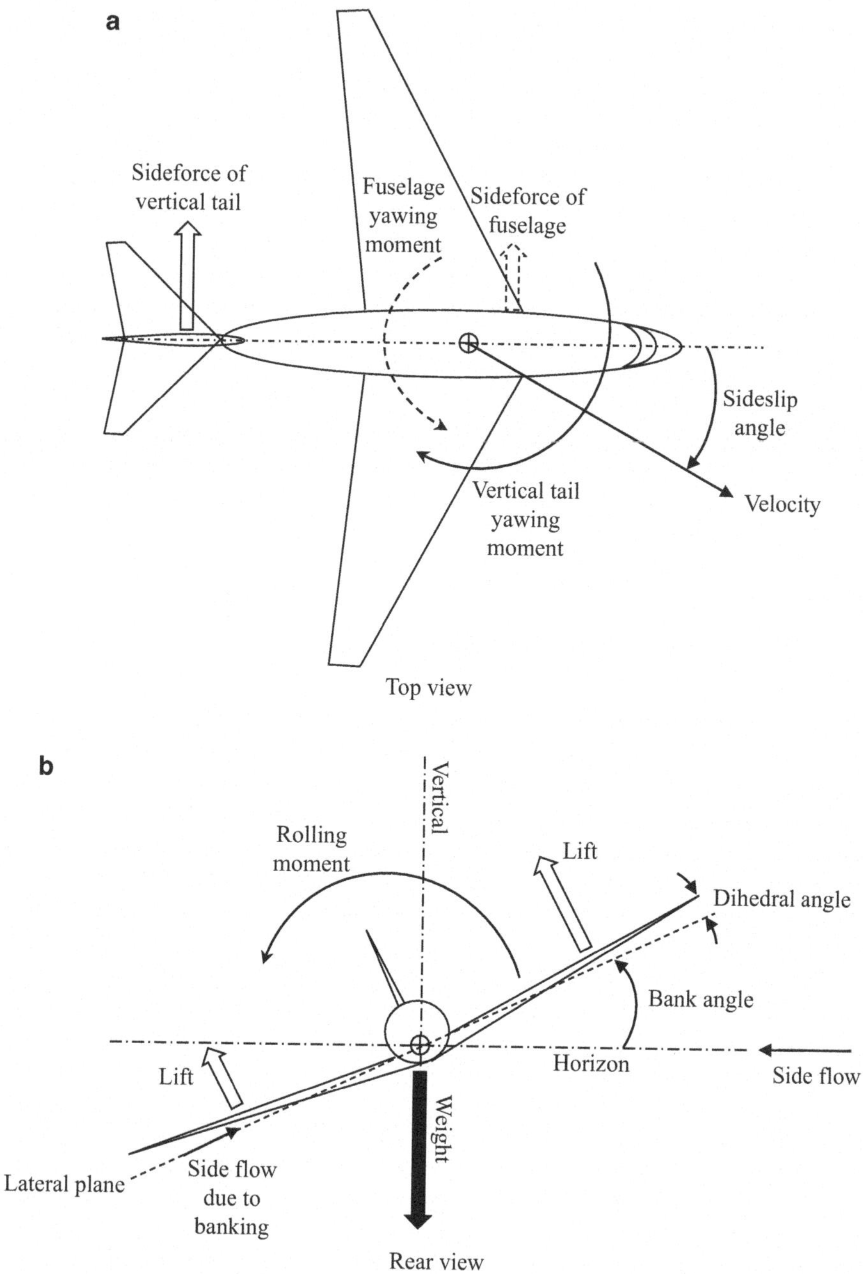

Fig. 4.3 Static stability in lateral–directional motion. (**a**) Directional static stability (or the weathervane effect) is produced by the vertical tail, which generates an opposite yawing moment any time a sideslip is encountered, thereby tending to restore the equilibrium condition of the fuselage pointing in the flight direction. (**b**) Lateral static stability (or the dihedral effect) is produced by the wings due to the side flow creating a differential lift on the left and the right sides. The rolling moment generated in this way banks the airplane such that it sideslips due to gravity in the direction opposite to the disturbance, thereby decreasing the net sideslip angle

in the lift produced by the left and right parts of the wing generates a rolling moment such that the aircraft banks away from the direction of the gust (in this case, toward the left). The bank angle makes the aircraft slideslip toward the left due to the lateral gravity component now acting along that direction, and the net sideslip angle is thus decreased. Such a natural static stability in roll is called the *dihedral effect*, because it increases as the *dihedral angle* of the wings is increased. The dihedral angle is the angle made by the wings with the lateral plane (i.e., the plane normal to the plane of symmetry) (see Fig. 4.3b). Since the lateral plane is initially parallel to the horizon in the equilibrium condition, any side flow would have a natural normal component to the wing due to the dihedral angle. For a wing mounted high on the fuselage, the dihedral effect is much larger than that for a low wing. Furthermore, a sweptback wing has a larger dihedral effect than a straight wing.

In the present discussion, we have neglected the "higher-order effects" of the lift of the fuselage, the change in the sideslip angle of the vertical tail caused by the wing's sidewash, and the rolling moment created by the vertical tail. Such effects do not detract from the main stability characteristics presented here, but are necessary for detailed calculations.

4.5 Dynamic Lateral–Directional Stability

In addition to the static stability of sideslipping, rolling, and yawing motions discussed above, all airplanes and gliders must have dynamic stability in these degrees-of-freedom. This is important both for the comfort of the passengers and crew in a transport aircraft, and for aiming and weapons delivery in military aircraft. An aircraft that goes on oscillating sideways, while rolling, and yawing, every time a sideways gust is encountered would be unacceptable as a suitable vehicle. For dynamic stability, it is important that such oscillations should be quickly damped out.

The damping in the yawing motion is provided mainly by the vertical tail, in a manner similar to the damping in pitch by the horizontal tail (or canard). As soon as a yaw rate is present in a particular direction (either nose left, or nose right), a sideforce is immediately generated by the vertical tail in the opposite direction. Since the sideforce is proportional to the change in the sideslip angle at the vertical tail, which itself varies directly with the yaw rate, there is an opposing yawing moment produced about the center of mass that is proportional in magnitude to the yaw rate. A larger yaw rate creates a larger opposing moment from the tail, and hence a natural damping is applied to the oscillations in the yaw angle. When the damping produced by the vertical tail is insufficient, an *active yaw damper* can be designed for the aircraft in using the rudder deflection in a feedback loop from the sensed yaw rate (via a rate gyro). Such active yaw control systems are often required for supersonic military aircraft.

The roll damping is chiefly produced by the wing. As soon as a roll rate exists, the wing experiences an increase of angle-of-attack on its down-going side, proportional to the roll rate. An equal and opposite change of the angle-of-attack is experienced at every point on the up-coming side of the wing. This creates a differential lift, which applies an opposite rolling moment in proportion to the roll rate, thereby dampening the roll dynamics. Due to its larger effective arm for the rolling moment from the longitudinal (roll) axis, a wing of a larger span has a much higher damping in roll compared to a smaller spanned wing of the same planform area. The increased resistance to the roll rate is also the reason why large transport airplanes take a longer time to roll by a given angle, when compared to a small fighter airplane, and are thus less maneuverable.

The roll rate also produces a yawing moment on the aircraft. This can be understood from the fact that wing will have a higher drag (due to its larger angle-of-attack) on its down-going side than on the one coming upward. The net differential drag thus applies a yawing moment in an *opposite* direction to that required for making a coordinated turn at the prevailing bank angle. This is called an *adverse yaw*, and is produced every time a roll rate exists, or when the ailerons are deflected. Such a yawing moment displaces the nose of the aircraft laterally, while the airplane is rolling. Aerobatic pilots take advantage of this natural adverse yaw by performing a maneuver called the *barrel roll* in which the airplane continuously rolls by 360°, describing a vertical circle tangential to the wing plane. A rudder input is also required in addition to the ailerons, to make the flight path follow the cylindrical shape of a barrel with an almost horizontal axis.

The cross-coupling existing among the sideslip, roll, and yaw degrees-of-freedom does not allow a disturbance in any one of the rates to remain isolated to the particular degree-of-freedom. For example, a sideslip disturbance applied suddenly at some time creates not only a sideforce, but also the rolling and yawing moments (as shown in Fig. 4.3). However, as soon as these moments are produced, the vehicle begins to rotate and translate, and the sideslip, roll, and yaw angles start changing with the time. Consequently, it is not possible to talk of rolling, sideslipping, or yawing dynamics in isolation. As in the longitudinal dynamics considered in Chap. 3, there exist various modes of later–directional dynamics, and each of them must be examined for a sufficient damping.

The simplest lateral–directional dynamic mode is the *pure rolling motion* excited by a sudden aileron input, in which the wing naturally acts as the dampening device. The damping in this mode is reasonably large, and thus the pure rolling mode is dynamically quite stable.

A mode in which the bank angle and the yaw angle are coupled together, whereas the sideslip angle is nearly zero, is called the *spiral mode*. A stable spiral mode results in a vehicle ultimately returning to the wing in the level position upon encountering a disturbance in the bank angle. In contrast, an unstable spiral mode consists of a slow divergence in both the bank (roll) and the heading (yaw) angles, such that aircraft enters an ever steepening, nearly coordinated turn. The flight path in the unstable spiral mode resembles a tightening spiral with the time, thereby giving its name. This mode does not involve any oscillations with time, but either a growth or a decay of the two angles. Being a long-period mode, it is easily controlled by the pilot when unstable. The spiral mode's stability depends mainly on the dihedral effect as well as the directional stability (the weathervane effect).

The spiral mode is usually produced by an aileron input applied in the level flight condition. This creates an initially constant bank angle, and a sideslip in the direction of the lowered wing due to gravity. This happens because a corresponding rudder input to maintain a coordinated turn has not been applied by the pilot. An aircraft that has an insufficient dihedral effect, tries to correct the sideslip by applying only a small rolling moment in the opposite direction. If at the same time, the static directional stability provided by the vertical tail is large enough to yaw the fuselage to cancel the sideslip, a mismatch would then exist between the rolling and yawing moments. While a zero sideslip condition will be achieved by yawing, the yaw rate thus produced would be larger than the coordinated turn rate. To achieve a balance of the normal forces, the bank angle would therefore continue to increase as the turn progresses, thereby causing an increased centripetal acceleration, and resulting in an unstable spiral mode. If the aircraft has an adequate dihedral effect in proportion to its directional stability, any change in the bank angle would tend to be corrected by a large rolling moment, resulting in an ever diminishing yaw rate, and thus a stable spiral mode. Clearly, spiral stability requires a large dihedral effect.

The third lateral–directional mode called the *Dutch roll mode* is a combined oscillation in the bank angle, the sideslip angle, and the yaw angle such that the sideslip angle is almost 180° out of phase with the yaw angle. This produces a snake-like motion of the airplane, rolling and yawing from side-to-side, such that the bank angle is always against the direction of the sideslip. The period of the oscillation is quite small (similar to the longitudinal short-period mode), but the oscillations could persist for some time if not properly damped. Since Dutch roll motion can be quite uncomfortable for the occupants of the aircraft, it is necessary that its damping should be as large as possible.

The damping in Dutch roll depends upon the dihedral effect, the damping in yaw, and the change in the yawing moment caused by the roll rate. For an aircraft with a large dihedral effect, the magnitude of the rolling moment produced by the sideslip is large, and thus feeds into the rolling oscillation due to that in the sideslip angle. Hence, there is a strong, natural feedback mechanism between roll and sideslip by a large dihedral angle, which tends to sustain the Dutch roll mode, decreasing its damping. For this reason, almost all passenger airplanes sacrifice the stability in the spiral mode in order to achieve an adequate Dutch roll damping by reducing the dihedral effect. This is carried out either by making the dihedral angle negative (called the *anhedral*), or by placing the wing much lower on the fuselage.

4.6 Lateral–Directional Control

Ailerons are the primary roll control surfaces of an airplane or glider, while the *rudder* is used to control the sideslip and yaw. Figure 4.4 shows how these surfaces generate their effective moments. While the ailerons are the primary devices for creating a rolling moment, they also produce an adverse yawing moment in a direction opposite to that required for a coordinated turn. This is due to the difference in the lift-induced drag on the two sides of the wing, which is created

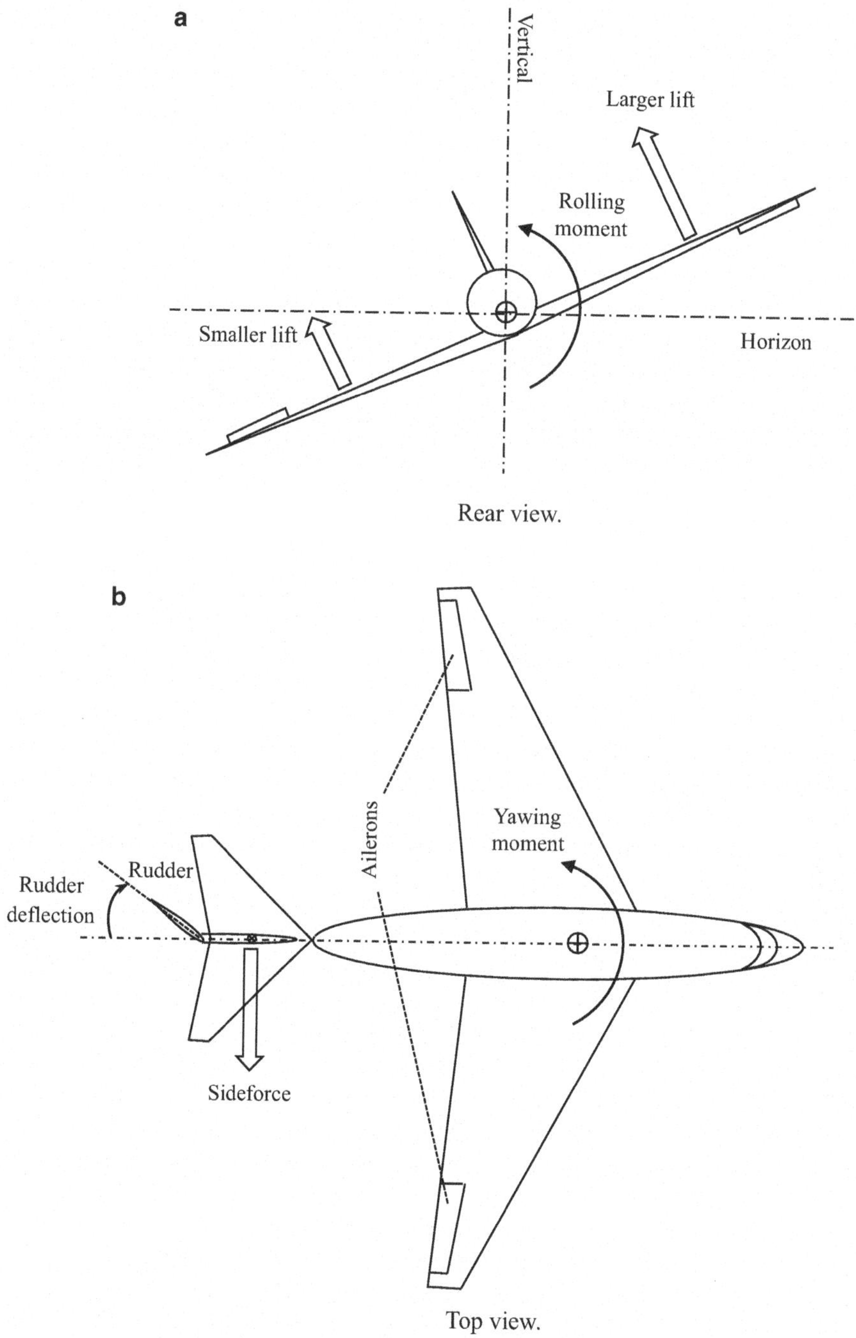

Fig. 4.4 Control surfaces for the lateral–directional motion of an airplane or glider. (**a**) Lateral control is mainly provided by the ailerons, which generate a rolling moment by creating a differential lift on the wing. (**b**) Directional control is provided by the rudder, which creates both a sideforce and a yawing moment by its deflection

when the ailerons are deflected. In order to decrease the magnitude of the adverse aileron yaw, both the ailerons are not deflected by the same angle. The up-going aileron is designed to deflect by a larger angle than the down-going aileron (see Fig. 4.4a), such that there is an increased parasite drag on the side of the wing which rolls down. In this manner, a part of the adverse yaw is canceled without affecting the lift (thus conserving the ailerons' effectiveness in producing the rolling moment). Some aircraft use a special cambered airfoil shape and an offset hinge-line for the ailerons (called *Frise aileron*) in order to alleviate their adverse yaw.

In order to balance the adverse aileron yaw, a rudder deflection in the direction of the turn is always necessary while initiating a turn. If this is not done, a sideslip will result due to the adverse yawing moment, which prevents a coordinated turn. As soon as the required bank angle for the desired turn rate has been achieved, the aileron deflection is maintained to hold the bank angle, while the rudder is brought to neutral. This is because in a coordinated turn already established, there is no need to create any further rolling moment, and thus no adverse yawing moment needs to be canceled by the rudder. An opposite aileron and rudder input is required again when coming out of a turn.

The rudder is quite effective in creating both a sideforce and a yawing moment, and is thus an important lateral–directional control device. While its use in a coordinated turn has been already mentioned, the rudder is employed every time an asymmetrical flight situation needs to be corrected. An example of this situation is a take-off or landing in the presence of a wind inclined to the runway. The aircraft must maintain a straight-line motion on the runway, otherwise a crash would follow. If the rudder is not applied in such a situation, the cross-wind velocity component acting on the fuselage and the vertical tail makes the aircraft veer from the runway. Another important use of the rudder is when an engine has failed during the take-off of a multi-engine airplane. The remaining engines would apply a yawing moment on the airplane, which if uncorrected by the rudder, would again cause it to veer from the runway. Furthermore, many glider pilots use the rudder for making a steep descending final approach for a landing whenever they find themselves too high (or too fast) for a safe landing. Since a glider does not have any engines, it has only one chance to make a good landing. If a landing is made from too high a final altitude (or too large a speed) at the runway threshold, there is a possibility of the glider overshooting the runway. A use of spoilers (see Chap. 3) normally decreases both the speed and the altitude quickly in such a situation, but a descent with spoilers extended is often too steep for many pilots' liking. Therefore, a maneuver called the *forward slip* is often employed, in which the rudder and aileron inputs applied in the opposite directions maintain nearly constant sideslip and bank angles. The glider is thus flying in a straight line with the fuselage yawed to one side, and the wing at an opposite bank angle from that required for a coordinated turn. The increased drag of the fuselage due to the sideslip can be easily controlled by the rudder deflection in order to produce a desired descent rate. Thus the pilot has complete control on the angle of descent, while also keeping the runway in view due to the forward banked attitude of the aircraft.

The use of the rudder is also indispensable in spin recovery, as explained next.

4.7 Spin and Recovery

An airplane or glider can experience a departure from controlled flight during an extreme maneuver called a *spin*. The spin is a coupled longitudinal and lateral–directional motion of an aircraft in a stalled condition, which involves a rapid rotation about pitch, roll, and yaw as the airplane descend almost vertically. Since a spin always results in a rapid altitude loss, and its recovery may put high stresses on the structure, it is a dangerous flight situation and must be avoided.

Figure 4.5 shows an aircraft undergoing a spin. The wing—being in a stalled condition—is producing a negligible lift, while the drag and the sideforce acting on the wing, the fuselage, and the vertical tail apply yawing and rolling moments to sustain a nearly constant lateral–directional rotation rate, called the *spin rate*. The horizontal tail could be producing some lift in a spin, and can cause a pitching oscillation resulting in the nose of the aircraft bobbing up and down as it spins. As the aircraft spins, it travels almost vertically downward with a constant speed produced by the weight nearly balanced by the net drag (which acts upward).

The spin characteristics greatly depend upon the mass distribution, i.e., the moments of inertia about the roll, pitch, and yaw axes, because the aircraft

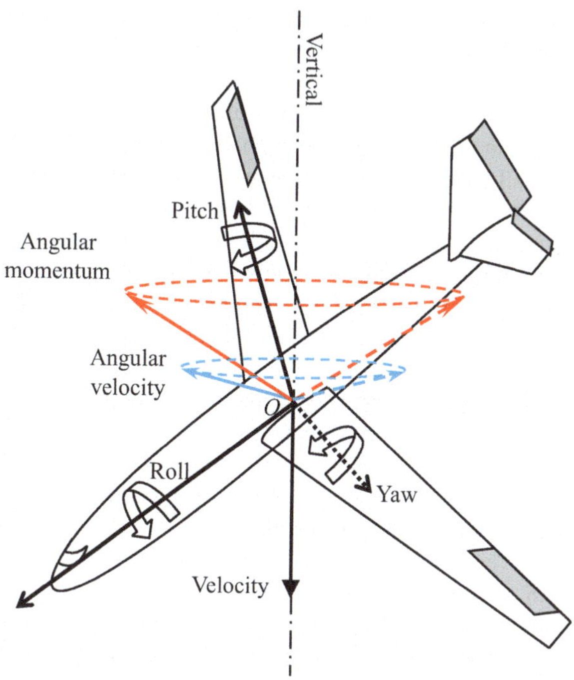

Fig. 4.5 An aircraft in a spin. The angular velocity and angular momentum vectors describe vertical cones due to the roll, pitch, and yaw rotations. The center of mass, *O*, travels almost vertically downward, resulting in a rapid loss of altitude at a nearly steady velocity

undergoes a multi-axis rotation at nearly a constant rate (magnitude of the angular velocity). We recall from Chap. 1 that the angular momentum vector in a multi-axis rotation is a product of the inertia tensor (which we left undefined) and the angular velocity vector. An external torque due to drag and sideforce on the various components continuously changes the direction of the angular momentum vector, causing it to describe a vertical cone with vertex at the center of mass. Similarly, the angular velocity vector also describes a cone about the vertical (see Fig. 4.5). However, the magnitudes of both the angular momentum and angular velocity vectors (i.e., the sides of their respective cones) remain nearly constant, producing a constant spin rate.

Another way to understand the spin dynamics is by energy considerations. As the spinning motion is "fueled" by the loss of altitude due to gravity, the energy lost by descending is converted into the rotational energy of the aircraft, and thus sustains its constant spin rate. Because it is not required to apply pilot inputs for keeping an airplane in a spin, it is also termed an *auto-rotation* of the aircraft (much like the windmilling rotor blades of a helicopter when it descends after encountering an engine failure). However, stopping a spin—called the *spin recovery*—does require deliberate actions by the pilot.

For spin recovery, a rudder input must be applied to generate a yawing moment against the spin direction. As the yaw rate is thus nullified, a normal stall recovery can be made by decreasing the angle-of-attack of the wing by elevator input. This kind of recovery works for most aircraft and is called the *full-opposite rudder, full stick forward* technique.

The most dangerous spin is the one in which nearly a horizontal attitude of the aircraft is maintained. This is called a *flat spin*, and is often experienced by airplanes with a small wing span. The moment of inertia about the yaw axis being predominant, a yawing rotation results in the smallest rotational energy for a given angular momentum, and is thus a natural condition of auto-rotation. An aircraft in a flat spin is yawing nose-to-tail, with a negligible rolling and pitching motions. This makes the sideslip angle at the vertical tail so large that it stalls, and therefore fails to generate a yawing moment when the opposite rudder is applied. A flat spin almost always results in a crash, and every aircraft is designed to avoid such a condition.

Any aircraft can be made to spin by applying a rudder or aileron input as it stalls. The spin will proceed in the direction of the applied pilot input. However, the spin is the most dangerous when it is inadvertently encountered during a normal maneuvering of the airplane close to the ground. Therefore, every aircraft must be designed such that it does not enter a spin during a normal stall.

The tendency to enter a spin is an important part of an aircraft's stalling characteristics. An aircraft which has a wing dropping to one side while stalling is much more prone to enter an inadvertent spin. The tip stall condition, where the flow separation takes place first at the wing tips, and then proceeds to the root as the wing's angle-of-attack is increased, is an ingredient for a spin. Every high aspect-ratio wing is therefore built with a twist angle in which the geometric angle-of-attack at the tips is made deliberately smaller than that at the root. This is called a *wash-out* design feature. Other wing design features, such as the taper ratio and the sweep angle may also determine the stalling characteristics.

Chapter 5
Flapping and Rotary Wing Flight

5.1 Introduction

In the previous chapters, we saw the flight of airplanes and gliders requires that a downwash must be created by the fixed wing as it moves through the air. The aerodynamic lift thus created depends crucially upon the speed relative to the atmosphere (airspeed), as well as on the incidence angle (angle-of-attack) made by the flight direction with the chord of the wing. As the speed falls below a certain level (called stall speed), the lift is insufficient to support the weight of the vehicle at even the largest possible angle-of-attack, and a stall results. Therefore, the fixed-wing flight requires a certain minimum forward flight velocity called the stalling speed. This severe limitation of airplanes and gliders is removed by having either a flapping wing or a rotary wing, which can generate the lift by the motion of a wing relative to the vehicle. In this manner, the vehicle can either ascend of descend vertically, or even remain stationary in the air (hover) like a hummingbird. In other words, while an airplane gently coaxes the air to follow the contours of its fixed wings, a flapping vehicle and a helicopter beat the air into submission with their moving wings.

The earliest ideas of flight (such as the "Ornithopter" of Leonardo da Vinci) were based upon the natural flight of the birds, which required a wing that could flap in an up and down motion relative to the flight direction. If the frequency at which the flapping is performed becomes large, it was assumed that the aerodynamic lift so generated would balance the weight, thereby enabling flight. Unfortunately, the aerodynamic principles preclude such a simple lifting mechanism, and even now, after more than a century of powered flight, there is no machine available to propel and lift human beings in this manner. The complex flow patterns generated by a flapping wing remain to be fully understood, largely due to the unsteady and partially separated flows involved.

A. Tewari, *Basic Flight Mechanics*, DOI 10.1007/978-3-319-30022-1_5

A rotorcraft such as a helicopter beats down the air with a large rotor to create the aerodynamic lift. While this principle is essentially the same as that of the flapping-wing flight, the flow pattern created by a rotor is quite different from that of a flapping wing. Furthermore, the relative simplicity of the rotor compared to a flapping wing makes it much more practical and reliable. That is why helicopters are ubiquitous, but ornithopters are nowhere to be found in the real world.

5.2 Why We Can't Fly Like the Birds

One of the earliest dreams of man is to gracefully fly like a bird, soaring into the skies and alighting upon a tree or a mountain at will. However, this dream has never been realized. The birds have powerful muscles and a very light body compared to the other animals. Their light feathers and bones are just sufficient strong to support the flight dynamic stresses, as well as to house the vital organs. The muscular strength of birds is several times larger in ratio to their body weight, when compared to that of humans. Furthermore, their wings are very flexible and can instantly morph into a variety of shapes as required. If we attach bird-like wings to our arms and try to flap them, we can neither have the same flexibility, nor the muscular strength to achieve flight, because the amount of air displaced by our flapping cannot generate sufficient amounts of lift and thrust. In fact, the flapping motion of a rigid surface increases its drag tremendously due to flow separation. Birds can manage to fly with a minimal flow separation due to the morphing of their wings, which adjusts the local angle-of-attack at any given point on the wing. It may thus be said that the birds are optimized for flight, while we are not.

5.2.1 Flapping Flight Basics

The simplest way of understanding the lift generation by flapping is by Newton's third law of motion. By a downward flapping stroke called the *plunge*, the air flow is turned downward and applies an equal and opposite (upward) reaction to the wing, which is lift. However, if there is no change in the angle made by the wing with the flight direction (i.e., the geometric angle-of-attack, or the *pitch* angle) during the upstroke (or the *heave*), the wing will experience a negative lift while moving upward; hence the net lift per flapping cycle would be zero. Similarly, the plunging–heaving motion alone cannot generate a net forward force per cycle on the aircraft (the *thrust*), because there is no possibility of turning the lift component forward. A sailboat can turn tack into the wind by turning its sail such that a component of the "lift" points against the wind. Similarly, the wing must also rotate (pitch) continuously while flapping such that it has an optimum angle-of-attack at all times to maximize the net lift and thrust on the aircraft per flapping cycle. This general principle behind the flapping-wing flight is shown in Fig. 5.1. The presence of the

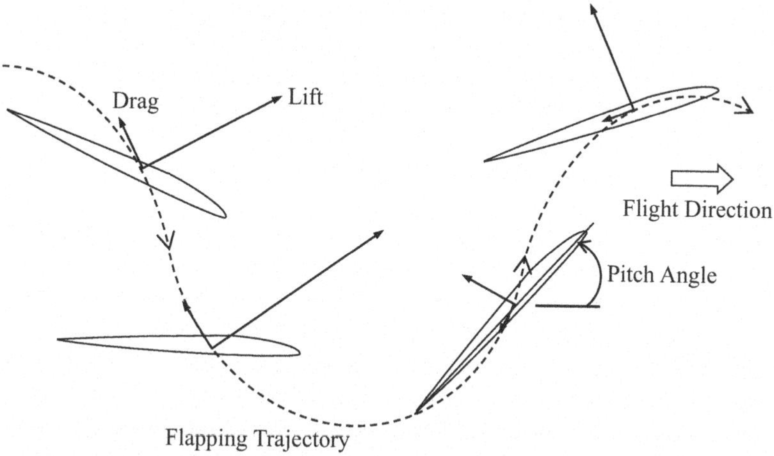

Fig. 5.1 General principle behind the flapping-wing flight: modulating a combination of plunging and pitching motions relative to the flight direction produces an ever changing lift and drag at each airfoil section, whose components are resolved as the net lift and thrust of the aircraft. (The pitching moment about the aerodynamic center is not shown here for simplicity.)

drag further complicates matters, because it always saps energy. An ideal flapping mechanism is the one which maximizes not the lift of the airfoil section, but the lift-to-drag ratio per cycle. The conversion of the local lift and drag into a net lift and thrust for the aircraft also necessitates a continuously modulated pitch angle. It would help tremendously if the wing were not a rigid structure, but could also change its shape (i.e., *morph*) by either curling into, or away from the flight direction during each stroke. The continuous pitching and morphing of the wings should be such that both a sufficiently large angle-of-attack is maintained and an excessive flow separation is avoided at all times. The birds are able to do this naturally and instinctively, while a machine would require an extremely complicated actuating mechanism and control logic to perform the same task.

Most creatures can swim in the water because their body weight is very nearly supported by buoyancy (see Chap. 1), thus only a little additional "lift" is required by displacing water through the swimming strokes. However, even this little dynamic lift must be produced by an optimum combination of strokes that involve a continuous twisting, downward and backward motions. Anybody trying to learn swimming quickly adapts from a wild splashing and thrashing motion, to the smoothly synchronized strokes of the arms and the legs. The birds and insects similarly produce an optimum flapping motion which is synchronized to achieve the ideal flight efficiency. Since there is a negligible buoyancy from the air to support a heavier-than-air flight, nearly all the lift must be produced dynamically.

Rather than the flapping frequency, it is actually better to talk of a *reduced frequency* (also called the *Strouhal number*), which is defined as the flapping frequency multiplied by the characteristic length, and divided by the speed of the flow. The Strouhal number roughly indicates the frequency necessary to vertically displace a given volumetric mass of air per cycle of the flapping motion. If we had proportionally the same wing area relative to our weight as the birds, our wings will be much larger; therefore we would have to flap our wings much more slowly compared to the birds in order to match their Strouhal number. However, the power required for flapping such large wings would be much higher than what could be provided by our muscles.

The attempts to build powered airplanes on the flapping lift principle have hitherto failed (except for very small experimental drones) due to the unavailability of very high power-to-weight ratio engines. While actively morphing wings that optimize the local angle-of-attack for every flight condition can be theoretically designed, their fabrication requires a large number of actuators (electric motors and reciprocating devices), as well as a highly flexible yet strong structure. The weight of such a mechanism becomes too large to be lifted solely by flapping, unless the size of the vehicle is very small. How can the size of a vehicle determine its ability to fly?

We see that small insects can accelerate more quickly, and are able to maneuver much better in a tight space than the birds. This indicates that insects can generate a much higher normal acceleration, therefore have a larger lift-to-weight ratio than the birds. On examining insect flight, we find that insects have much smaller wings relative to their body dimensions when compared to that of a bird. Consequently, the Reynolds number of insect flight is several orders of magnitude smaller than that of a bird. From Chap. 2, we recall that the Reynolds number is the ratio of the flow's momentum per unit cross-sectional area to the flow's viscosity, and is proportional to the characteristic length. Since an insect's wing has a much smaller characteristic length, its Reynolds number is quite low, therefore the viscous forces predominate during its flight. The mechanism of lift creation in a highly viscous flow is completely different from that in a flow with a small viscosity. Apparently a much higher lift-to-drag ratio can be achieved at the very low Reynolds numbers characteristic of insect flight, thereby requiring a proportionally smaller power for a given lift.

This explain why the low Reynolds number flapping flight of micro-unmanned aerial vehicles (MAVs) is possible, but a manned flapping flight vehicle that has a much higher Reynolds number than an MAV requires a much higher flapping power, which is not feasible with the available technology.

5.2.2 *Basic Flapping Mechanism*

The most flapping mechanism is the one which allows a morphing of the wing's shape in both chordwise and spanwise directions. Consider a simple mechanism shown in Fig. 5.2 operated by two rigid links: *AB* and *BC*. Each link can assume a different angle entirely depending on the displacement of its ends. Suppose the link *AB* has its end *A* sliding horizontally along a guide in the body of the vehicle such

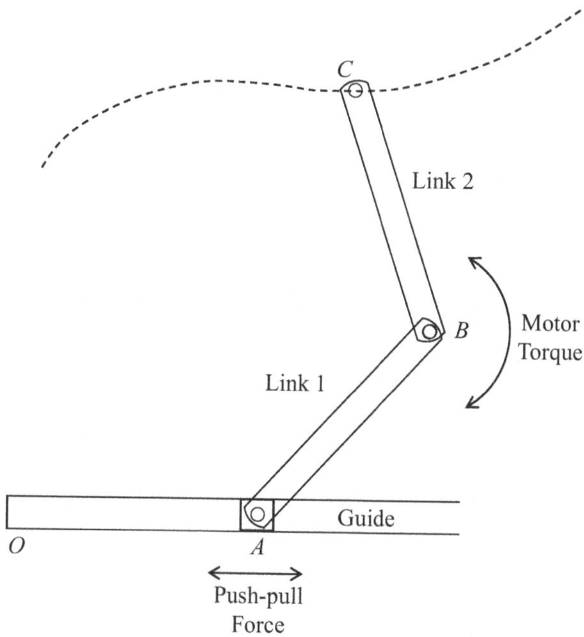

Fig. 5.2 A simple actuating mechanism for flapping and morphing at a given control point *C* on the wing's surface

that there can be no vertical displacement of *A*. Hence, the entire link *AB* can move horizontally at *A* as well as rotate about *A*. Such a compound motion can be driven by a reciprocating (rocker or push-pull) device, which applies a horizontal force at *A* and a motor which applies a torque about *A*, as shown in Fig. 5.2. Another motor placed at the intersection of the two links applies an independent torque about the point *B*. As a result, the point *C* describes a curve, whose shape depends upon the angles *OAB* and *ABC*, as well as the horizontal displacement of the point *A*. Hence both vertical and horizontal displacements of the point *C* are controlled by the two motors and the reciprocating device.[1]

If the point *C* is on the wing's surface, then this mechanism deflects the wing both vertically and horizontally by desired amounts at that point (called control point). Two (or more) such mechanisms placed at different locations can produce any desired complex shape of the wing's surface. However, the number of actuating devices (rockers and motors) required to generate such a shape would increase rapidly with the number of control points. For example, if both the chordwise and the spanwise shape were to be controlled, then a minimum of two such mechanisms would be necessary, which requires a total of six actuators. Each actuator must be driven by separate electrical signals and must have a different

[1] If the point *C* is fixed to the body, then this system becomes a simple *slider-crank* mechanism used in piston engines, where the linear (reciprocating) motion at *A* is converted into a rotary motion of the link *BC*, and vice versa, which can be driven by either the force at *A* or the torque at *B*.

control logic. The control logic must ensure not only the desired shape, but also structural compatibility of the displacements at every point, otherwise the wing can come apart. Thus it can be imagined that a flapping and morphing mechanism can become extremely complicated as the number of local actuating devices is increased. In a bird, the control points are numerous and the actuation is performed by muscles which are distributed in both spanwise and chordwise directions, rather than localized as rockers and motors. The signals to drive each muscle are generated by the brain and transmitted via nerves. By its very complexity, the bird-like flight is still beyond our capability.

5.3 Rotary Wing Flight

The aircraft which use a large rotor to create lift are called *rotorcraft*, and are divided into *helicopters* and *autogyros*. Like an airplane, a helicopter uses engine power to drive its rotor, whereas an autogyro does not have an engine, but its windmilling rotor is driven by the motion through the air powered by the gravity. For this reason, while a helicopter can hover and make vertical take-offs and landings, an autogyro must always lose altitude by flying at an angle to the vertical (like a glider). Let us briefly examine how a rotary wing flight takes place. Because an autogyro can be visualized as a helicopter with a failed engine, we will confine most of the discussion to helicopters.

The flight of helicopters and autogyros uses the downwash created by a large rotor for producing the aerodynamic lift. By Newton's third law of motion (see Chap. 1), the downward deflection of the air passing through the plane of the rotor generates an opposite reaction in the upward direction, which is the lift. Figure 5.3 shows the relative airflow created in the vicinity of the main rotor in a vertical flight (ascent, descent, or hover). The streamlines enclosing the rotor have a converging area between them, which indicates an increase in the speed of the flow as it approaches and passes the rotor. It is this downward rate of increase of the momentum of the relative airflow which generates the lift by Newton's second law. Because the fuselage is almost completely submerged in the relative flow produced by the main rotor, the fuselage drag always acts downward in a vertical flight and must be subtracted from the rotor's lift to give the net lift on the vehicle.

When the rotor is tilted in a horizontal direction, a component of the rotor's lift acts as the thrust, causing the vehicle to move in that direction. A drag component of the fuselage now opposes the horizontal motion and acts as the drag on the vehicle. The vertical component of the total lift produced by the rotor is the net lift on the vehicle, and must balance the weight for equilibrium. This implies that the rotor's lift in a forward flight must be increased from its value in the vertical flight. The forward flight of a helicopter at a steady and level condition is schematically depicted by Fig. 5.4. The turning of the relative airflow in the downward direction and the increase of its speed by the rotor are evident in Fig. 5.4. An autogyro—being unpowered—is unable to maintain level flight, and is thus always descending. If an

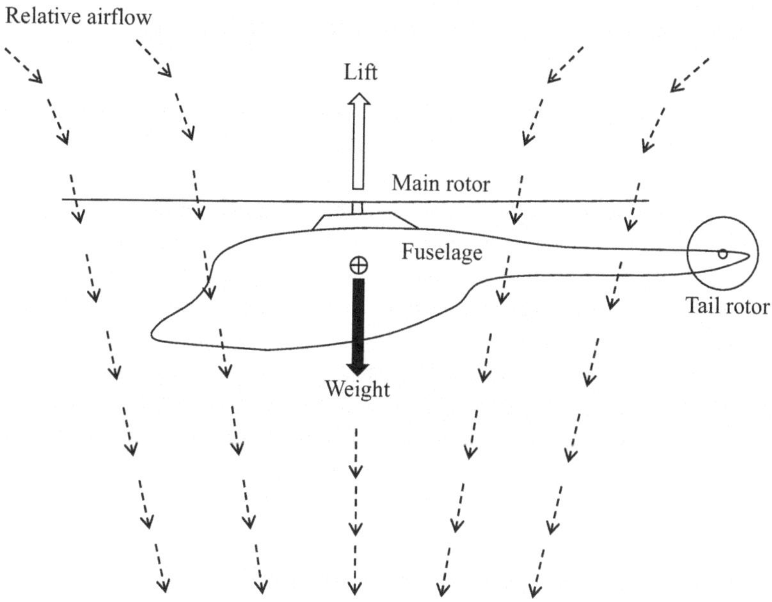

Fig. 5.3 The vertical flight of a helicopter. The net lift must balance the weight for an equilibrium

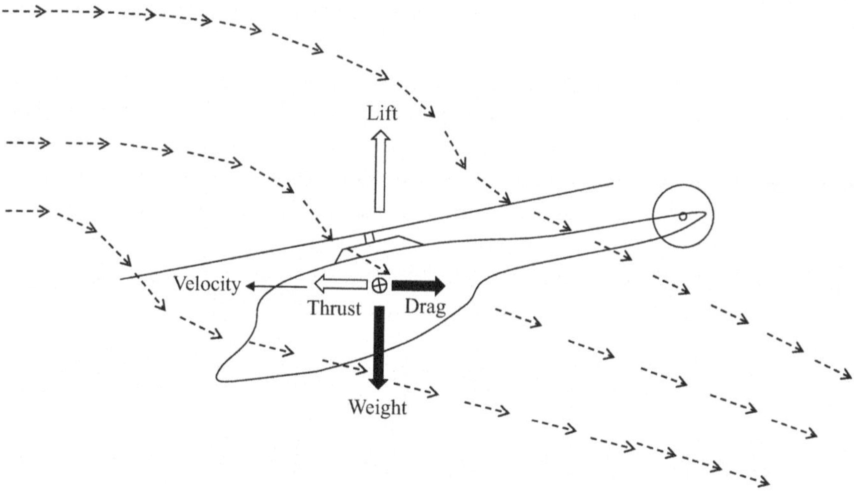

Fig. 5.4 The forward flight of a helicopter. The rotor's lift is resolved in a net lift and a net thrust on the vehicle for balancing the weight and the drag, respectively, for a steady and level flight

equilibrium of the forces is maintained, then an inclined descent at a constant angle can take place like a glider. The range and endurance of the autogyro entirely depend upon the rotor's efficiency in producing the lift.

The ability to take-off and land vertically, to hover, and to fly horizontally provides a high maneuverability to the helicopter, which is unmatched by any other vehicle. A helicopter can fly forward, backward, sideways, vertically up or down, and in any combination of these ways. Furthermore, it can also maneuver vertically and horizontally like an airplane (see Chap. 3).

If there were no losses, the engine power driving the rotor would exactly equal the rate of change of kinetic energy of the air deflected downward. However, because the rotor also imparts a rotary motion to the air, as well as due to the effects of viscosity and compressibility (see Chap. 2), a part of the engine power is lost in the useless rotation, friction, and compression of the fluid medium. The efficiency of a rotor is measured by the ratio of the lifting power produced to the total engine power supplied to the rotor by the engine. The design of the rotor must be based upon a minimization of these losses, such that the highest possible efficiency is achieved in a design flight condition. Furthermore, the fuselage must have a shape that creates the smallest possible drag in the given flight condition.

5.4 Vertical Ascent, Descent, and Hover

The principle of rotor lift is easily understood for vertically ascending (or descending) flight. Here, the airflow through the rotor is vertical (if we neglect the rotational airflow) as shown in Fig. 5.3. Far upstream of the rotor, the flow speed is the same as that of the helicopter. The airflow is accelerated as it approaches the rotor, and continues to speed up until reaching its maximum velocity far downstream. Consequently, the area of the airflow bounded by the streamlines continues to decreases, as shown in Fig. 5.3. The net acceleration of the flow downward produces the lift on the rotor as an equal and opposite reaction. This is the same principle by which the thrust is produced by a propeller. By having the lift exactly balancing the weight, the helicopter can remain in an equilibrium in a hover.

For vertical flight at a constant speed, the rotor's lift must be greater than the weight such that it equals the weight and the drag of the fuselage. Since the net relative airflow produced by the rotor is always in the downward direction for generating the lift, so is the drag caused by the fuselage. Thus the drag of the fuselage should be always subtracted from the rotor's lift in a vertical flight. (In a strict terminology, the rotor's lift in a vertical *descent* is actually a *drag* on the vehicle, because it acts against the downward direction of flight. In that sense, the fuselage drag during a steady descent can be called a *thrust* because it is produced along the flight direction (i.e., downward).)

5.5 The Conundrum of Forward Flight

While moving horizontally, the rotor must be inclined at an angle with respect to the vertical in order to produce a forward thrust plus the lift (see Fig. 5.4). Therefore, the net lift produced by the rotor should be greater than the weight of the helicopter.

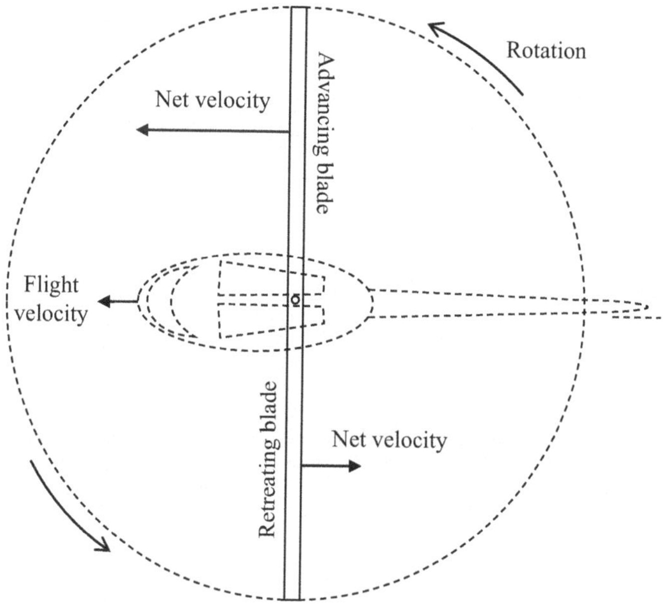

Fig. 5.5 The forward flight of a helicopter results in a rolling moment on the rotor due to the advancing and retreating blades seeing different net velocities, and the lift. If not taken into account and corrected, this will produce an undesirable roll rate on the vehicle, and a departure from the straight and level flight will follow

However, the very fact that the rotor has a forward velocity component causes a difficulty. As shown in Fig. 5.4, the plane of the rotor experiences a horizontal flow component (in addition to the vertical downwash required for the lift creation). Due to the horizontal airflow, the advancing blade of the rotor experiences a larger relative airspeed when compared to that of the retreating blade (see Fig. 5.5). This variation of the relative airspeed due to rotation causes a larger lift on the advancing blade, and a smaller lift on the retreating blade, thereby producing an undesirable rolling moment on the rotor. Since there is no way of counteracting a rolling moment in a helicopter, there would be a roll rate produced on the vehicle, which will immediately cause a departure from the straight-line forward flight. In an extreme case, the net speed reduction on the retreating blade can also cause a *retreating-blade stall*, which means a complete loss of lift in a portion of the rotating cycle.

In order to avoid the adverse rolling and the retreating-blade stall problems, the helicopter blades are *articulated*. This concept is shown in Fig. 5.6. Instead of being rigidly attached to the hub, the blades are hinged and restrained by springs. The hinging mechanism of each blade allows a flapping (up and down) motion, a pitching rotation, and a lead-lag motion in the plane of the rotation. These can be considered to be the degrees of freedom of the rotor as a rigid body. As the lift begins to increase on an advancing blade, it flaps about the hinge, resulting in an upward motion of the blade airfoil (called the *heave*). Furthermore, its increased

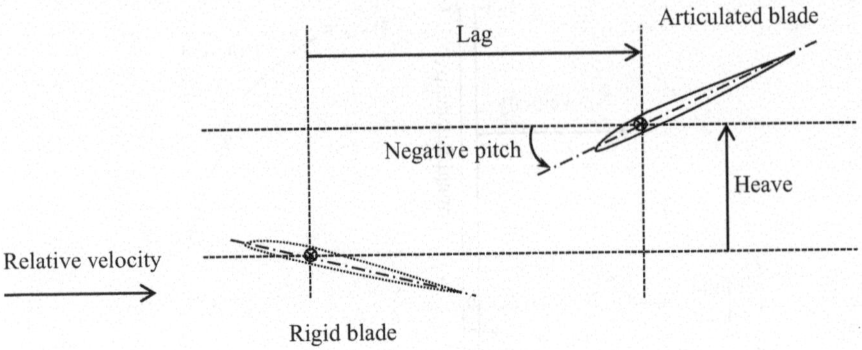

Fig. 5.6 Articulation of the advancing rotor blade results in a heave, lag, and negative pitch of the airfoil due to the increased relative flow speed. The retreating-blade airfoil has the opposite effect (plunge, lead, and positive pitch). Articulation allows a nearly constant angle-of-attack at each spanwise station of the blade through the rotation cycle. In addition, the aeroelastic effects cause a natural flapping and twisting of the blades, which aid the articulated hinge at the hub

drag also causes a lagging (retreating) motion, which results in a moment arm that applies a nose-down pitching moment about the pitch axis. The combined heave and negative pitch cause a decrease in the angle-of-attack (see Fig. 5.6) at all points on the advancing blade, thereby reducing its lift. In contrast, a retreating-blade airfoil seeing a tendency of the lift and drag to decrease, experiences the downward (plunging), leading (advancing), and positive (nose-up) pitching motions, thereby increasing its angle-of-attack and also the lift.

The favorable effect produced by the articulation of the blade is aided by the structural flexibility of the rotor blade, which flexes up or down depending upon the lift acting on it. It can also twist due to the torsion provided by the aerodynamic pitching moment. Hence, a natural aeroelastic feedback mechanism is established wherein the rigid (heave-plunge, lead-lag, and pitch) and the flexible (bending and twisting) degrees of freedom counteract any tendency of the lift to either increase or decrease. A stable situation is thus established on each blade of nearly a constant lift through the rotation cycle. It is the very flapping and the flexing of the blades as the helicopter flies forward which is responsible for the chopping sound produced by it (hence its popular name, the "chopper"). However, the articulation of the blades can prevent the adverse rolling and the retreating-blade stall only up to a certain forward speed, which imposes a maximum speed limit on a helicopter. Therefore, the forward flight speed of a helicopter is often determined by the articulation mechanism employed in the rotor.

5.6 Need for a Tail Rotor

When a rotor is put into motion by the engine, the rest of the vehicle (i.e., the fuselage) tries to rotate in the opposite direction due to the principle of the conservation of angular momentum (Chap. 1). If allowed to do so, this rotary motion of the fuselage would be extremely uncomfortable for the passengers and the crew. Furthermore, an additional loss of flight efficiency would result from the drag created by the sideways rotation of the fuselage. Therefore, to keep the fuselage in equilibrium a small tail rotor is used to cancel the sideways torque produced by the rotor. The addition of a tail rotor complicates helicopter design due to the requirement of a driving mechanism. It also creates a hazard while taking-off and landing due to the possibility of the tail rotor striking the ground. In order to avoid having a tail rotor, some helicopters have two contra-rotating rotors instead of a single main rotor. The rotors can either be coaxial (like in the *Kamov Ka-52*), or on different axes (like in the *Boeing CH-47 Chinook*). Since they are rotating in the opposite directions at the same rate, the contra-rotating rotors do not produce a net torque on the vehicle. Another option of doing away with the tail rotor is the concept of *no-tail-rotor* (or NOTAR) by having a sideways jet exhaust (or Coanda effect) at the tail for canceling the main rotor torque. So far, the NOTAR concept has been applied only by the McDonnell Douglas, such as on the *MD Explorer*.

5.7 Controls for Rotorcraft Flight

Controlling the flight of a rotorcraft requires an ability to apply a force on its center of mass to cause an acceleration in a specific direction. Furthermore, their must be a way of rotating the fuselage by applying rolling, pitching, and yawing moments. The lift of the main rotor can be used to generate a force in a particular direction. The pitch and roll motions can also be produced by the main rotor, while the tail rotor can apply a yawing moment.

In order to control the lift of the main rotor, it is necessary to have ways of changing the angle-of-attack of each rotor blade. The simplest way of doing so is by changing the pitch angle of each blade at the hub. This can be carried out in the following two ways.

A *collective* control applies the same constant, pitch angle to all the blades, thereby increasing (or decreasing) their lift by the same amount. In this manner, the magnitude of the rotor's lift can be adjusted as required.

The *cyclic* control continuously changes the pitch of each blade as it rotates, such that all the blades have the same pitch angle at a given point in the rotating cycle. This results in a tilting of the hub (along with the articulation hinges) in order that the rotation of each blade is carried out in a specific tilted plane. In this way, the rotor's lift can be inclined in any given direction, thereby producing a change in the flight direction.

Both collective and cyclic controls are often provided by the movement of the same control lever (called the *stick*) such that a sideways movement activates the cyclic resulting in a banking (rolling) motion, and the fore-and-aft movement changes the collective which results in a change in the rotor's lift.

In addition to the collective and cyclic controls, the speed of the tail rotor can also be adjusted to apply a torque about the rotor axis. This produces a net *yawing moment* on the fuselage which is used to align it in a desired direction. The torque control is activated by the *torque pedals*, such that pressing the left pedal results in a yaw to the left, etc.

The stability and control characteristics of a rotor require a knowledge of the unsteady aerodynamic and aeroelastic effects, which are beyond our present scope.

Chapter 6
Space Flight

6.1 Introduction

Space flight is the ultimate form of travel which takes humans from the confines of the earth, and space exploration is an important means of expanding the frontiers of science. However, there are many prevalent misconceptions regarding space flight, which are probably derived from the popular science fiction. One of them is the issue of weightlessness. Astronauts are supposed to become weightless as soon as they get into the space. This is far from being true. Our weight is the reaction we feel from the ground as we stand, sit, or lie on it, or from any platform which has a zero vertical acceleration relative to the ground. By Newton's third law of motion (see Chap. 1), if any object (such as yourself) exerts a force on any other object (like a platform), the second object also applies an equal and opposite force (called the reaction) on the first object. The magnitude of the force applied is equal to your own mass times the acceleration relative to the platform, according to the second law of motion (Chap. 1). If the platform itself does not have any vertical acceleration, the reaction experienced by you from the platform is equal to your mass times the acceleration due to gravity, which is your weight. Now, suppose you are in an elevator which is accelerating upward relative to the ground (see Fig. 6.1). The net acceleration experienced by you from the platform is now greater than that due to gravity, and it seems as if your weight has increased because the platform exerts a larger reaction on you. If a tragedy were to happen in which the elevator's cable snaps, then both you and the elevator would be falling down with the same acceleration due to gravity (this situation is called a *free fall*), therefore your relative acceleration due to the platform would be zero. Since now you will not feel any reaction from the platform, it would seem as if your weight has become zero. In fact, your weight is still the same as it was before you got into the elevator. Similarly, an astronaut only feels a zero reaction from the spacecraft, whereas both the astronaut and the spacecraft

A. Tewari, *Basic Flight Mechanics*, DOI 10.1007/978-3-319-30022-1_6

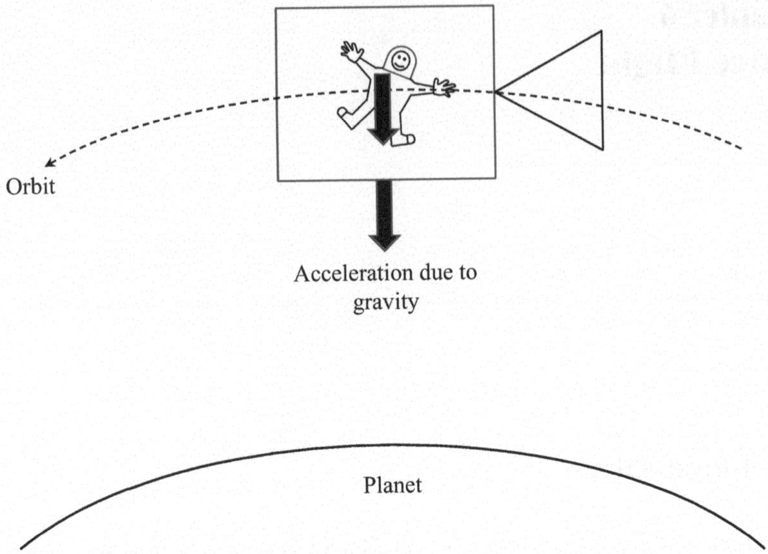

Fig. 6.1 The problem of weightlessness; a spacecraft and an astronaut are both moving with the same acceleration due to gravity, hence they do not apply any force (weight) on each other

are falling freely with the same acceleration due to gravity. The same weightless sensation is produced in an airplane which is making a vertical turn in a downward direction with a rate such that the lift acting on the aircraft is exactly zero, and the centripetal acceleration is provided only by the gravity. Such a zero-lift (or *ballistic*) flight is used to train astronauts for space missions.

The magnitude of the acceleration due to gravity is not a constant, but decreases with the inverse of the square of the increasing distance between the spacecraft and the center of the body causing the gravity (such as the earth, the moon, or the sun). When going to the top floor of a high building, while climbing a mountain, or even when flying in an airplane, the change in the vertical distance from the center of the earth is insignificant, hence the acceleration due to gravity (and your weight) is effectively the same as that at the sea level. However, the vertical distance changes significantly in the space, hence the magnitude of the gravity becomes smaller. From this discussion it may appear to be possible to escape the effect of a planet's gravity by going to a very large (infinite) distance from its center. But as one escapes the gravitational influence of a body (like a planet), one comes under the influence of the sun's gravity. Currently, two robotic spacecraft, namely *Voyager-I* and *Voyager-II* are on the verge of escaping the confines of the solar system. Far in the future, they might reach a point where the sun's gravity becomes insignificant. However, they would still be under the gravitational influence of our galaxy, the *Milky Way*, and sometime in the far future, they might also experience a tug from the gravity of another star! Therefore, it is never actually possible to become completely free from gravity (or weightless) at any point in the universe.

Other popular misconceptions have to do with the way a spacecraft is launched, and how it moves when it is in the space. Many people have in mind the picture of a rocket which lifts-off from the ground, and continues to fly vertically until it crosses the atmosphere into the space. When it reaches the space, the rocket is popularly supposed to become a spacecraft, traveling in any desired direction driven by its engines like a car or an aircraft. This chapter is basically written to remove such misconceptions regarding the space flight.

6.2 Orbits

Space flight is governed by the gravity of a massive body, in the vicinity of which the spacecraft is moving. Although the spacecraft is also being pulled by other bodies, the gravity of only one body (called the *primary* gravity) is significant during a selected part of its flight. This is because of the vast separation of the bodies in the solar system, and also the large differences in their masses. Newton's law of gravitation (Chap. 1) states that the gravity of a body is directly proportional to its mass, and inversely proportional to the distance from its center to the object (which is considered to be a particle) being attracted by it. This implies that a spacecraft orbiting a much more massive body (like the sun) experiences only a very small tug from a smaller body (like a planet), unless it happens to come very close to it. Furthermore, when the smaller bodies (such as the planets) are orbiting the larger bodies (such as the sun), a spacecraft in the vicinity of a smaller body roughly experiences the same acceleration due to the much more massive (but faraway) third body. Therefore, its relative motion around the smaller body is largely unaffected by the third body. Except in some special situations, it is usually a good assumption to treat the spacecraft as if it is being pulled by only one massive object (the primary body) during a significant part of its flight. This is termed the *two-body problem* of space flight. All the other bodies can exert a much smaller gravity (except under special circumstances) than the primary gravity. Hence, the translational motion of the spacecraft is generally described as an *orbit* around the primary body. An orbit could be modified over a long period of time by the small *perturbations* caused by the presence of the other bodies in the solar system, as well as due to other effects (to be explained later).

Since the gravity of a spherical body always pulls the spacecraft directly toward its center, it causes the spacecraft to move in a fixed plane along an orbit whose shape, size, and orientation are constant with the time. An orbit has the shape of a *conic section*, that is, the shape produced when cutting a right-circular cone with a straight knife. Of course, the cut can be made in various ways, each producing a different shape of the orbit. These shapes can be classified as being circular, elliptical, parabolic, and hyperbolic, and depend solely upon the *energy* and the *angular momentum* with which a spacecraft is flying. The orbital energy of a spacecraft is determined by the distance of its center of mass from the center of the primary (called the *radius*), and the speed of the center of mass measured in

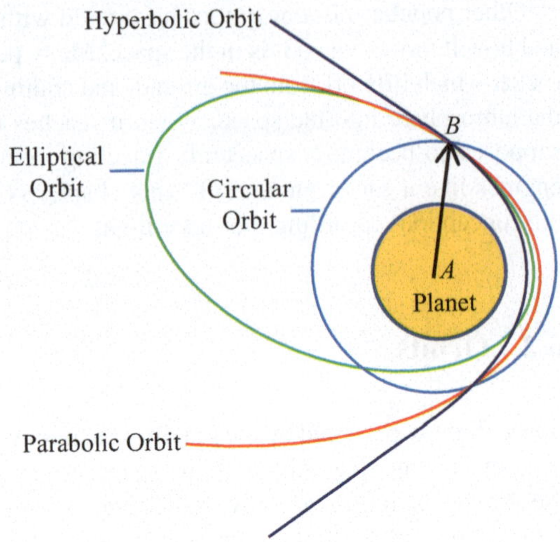

Fig. 6.2 Spacecraft orbits of various shapes sharing a common radius, *AB*, and plane: circular, elliptical, parabolic, and hyperbolic

an inertial frame. The inertial frame is conveniently located at the center of the primary, but does not rotate with the primary body. The orbital angular momentum is the angular momentum (see Chap. 1) of the spacecraft's center of mass about the primary center.

Figure 6.2 shows the various orbital shapes possible for a spacecraft around a spherical primary, beginning at a common radius—depicted as the arrow *AB* centered at the primary (*A*)—and pointing toward the current location of the spacecraft (*B*). Which shape of the orbit is actually produced depends upon the velocity vector (speed and direction of motion) at the given point, *B*. The velocity vector of the spacecraft is depicted as the arrow *BC* in Fig. 6.3. The simplest shape is a circular orbit, which is a special orbit whose radius is the same at all points. This requires that the radial component of the velocity vector must vanish at all points, which can only happen if the angle $\angle ABC$ made by the velocity vector with the radius vector is always 90°. If this angle differs from 90° at any point, then the orbit has a non-circular shape as shown in Fig. 6.3. In this case, the shape of the orbit (i.e., elliptical, parabolic, or hyperbolic) as well as its size depends only upon the lengths of *AB* and *CD*, as well as the value of the angle $\angle ABC$ between them, which determine the orbital energy and angular momentum.

For example, a satellite in a circular orbit of 6900 km radius (i.e., the length of the arrow *AB*) around the earth has a flight speed (i.e., the length of the arrow *BC*) of 7.6 km/s relative to a non-rotating earth, and is moving such that its velocity vector is always perpendicular to the radius vector. In contrast, a meteor approaching the earth at the same radius may have a velocity of 11 km/s that makes an angle of 120° with the radius vector. If this angle were 90° (as in case of the circular orbit), the angular momentum would be the largest for the given radius and speed. The higher energy and a correspondingly smaller angular momentum of the meteor's orbit translate into a hyperbolic orbit, which can have an unbounded (infinite) radius.

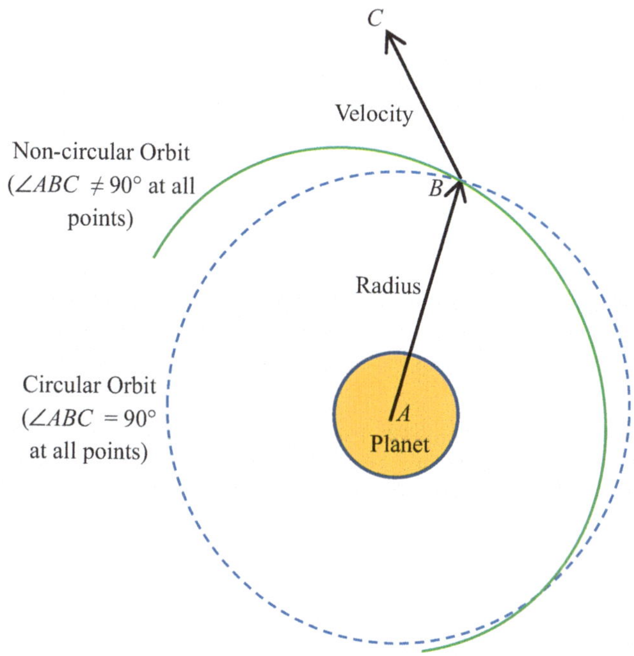

Fig. 6.3 Radius vector *AB*, velocity vector *BC*, and the angle $\angle ABC$ between them determine the shape of an orbit

Hence, a hyperbolic orbit is open orbit, and results in the spacecraft traveling almost in a straight line far away from the primary (Fig. 6.2). If the meteor had a much smaller velocity (say 7.4 km/s) which made the same 120° angle with the radius vector at the given radius of 6900 km, it would be captured around the planet in an elliptical orbit. Elliptical orbits have a much smaller energy compared to hyperbolic orbits, and remain bounded in radius as shown in Fig. 6.2. There are a point of the smallest radius and a point of the largest radius in an elliptical orbit, which are called the *periapsis* and *apoapsis*, respectively. For an orbit around the earth, the general suffix *-apsis* is replaced by a more specific *-gee* (i.e., perigee and apogee). Similarly, the corresponding points in an elliptical orbit around the sun are called *perihelion* and *aphelion*, respectively.

A parabolic orbit is the trajectory required to escape the primary gravity with the *minimum energy* at a given radius, and can be regarded as the boundary between elliptical and hyperbolic orbits. In the present example, if a spacecraft has a speed of 10.75 km/s when its radius is 6900 km, it will follow a parabolic escape trajectory, which will theoretically take it to an infinite radius from the earth, at which point it will come to a rest. The reason we have used *theoretically* as a caveat is that any spacecraft making a parabolic escape from the earth's gravity is immediately captured in an elliptical orbit around the sun, which of course, never lets it actually go to an infinite radius. If the energy is increased beyond that of parabolic escape, the spacecraft goes into a hyperbolic escape trajectory, which is such that the speed is non-zero even at the theoretically infinite radius. An interplanetary spacecraft has a hyperbolic orbit relative to the planets, but an elliptical orbit around the sun.

6.3 Orbital Maneuvers

Space flight often requires changing from one orbit to another. It is possible to change the shape, the size, and the plane of a spacecraft's orbit by applying an external force. Such changes are called *orbital maneuvers* and are usually performed by the use of rockets. Most spacecraft are equipped which rocket engines (see Chap. 7 for the principle of a rocket engine) that are used for making the orbital changes. Each time a rocket is fired, it changes the velocity vector of the spacecraft, resulting in modified energy and angular momentum vector. However, firing of rocket engines expends the precious propellants, therefore spacecraft orbits must be designed in order that the minimum amount of firings (thus propellants) is necessary.

6.3.1 Rocket Powered Maneuvers

The choice of an efficient rocket engine is crucial in orbital maneuvers. Rocket engines can apply a large thrust in a small duration, which results in an almost instantaneous change in the velocity vector. Such a change in the orbit is called an *impulsive orbital transfer*, and is rather like the change in a baseball's trajectory produced by hitting it with a bat. Figure 6.4 shows an impulsive orbital transfer where a single velocity impulse (the arrow CC') applied at radius AB instantaneously changes the shape, size, and the plane of the orbit. The magnitude of the net velocity change CC' is directly proportional to the speed of the exhaust gases ejected from the rocket engine, which is a constant for a given propellant and nozzle geometry (see below). The constant speed of the rocket exhaust divided by the acceleration due to gravity on the earth's surface is called the *specific impulse* of the rocket engine. The larger the specific impulse of a rocket engine, the smaller would be its propellant requirement for carrying out an orbital maneuver. For example, an engine employing a solid propellant has a specific impulse of 300 s, while a cryogenic engine using liquid hydrogen and liquid oxygen as the propellant can have a specific impulse of 460 s. An alternative to the chemical rockets that have been conventionally employed in space flight, are the plasma (or ion) propulsion engines which accelerate charged gas particles via a magnetic field. Such engines can currently produce only a very small thrust, but they have a much higher specific impulse (about 5000 s). Plasma engines cannot produce impulsive orbital changes (due to their small thrust), but can be used to make an orbit continuously evolve with the time. Such a maneuver is called a *low-thrust orbital transfer* and involves a slowly varying orbital shape, size, or plane. Some spacecraft such as the *Deep*

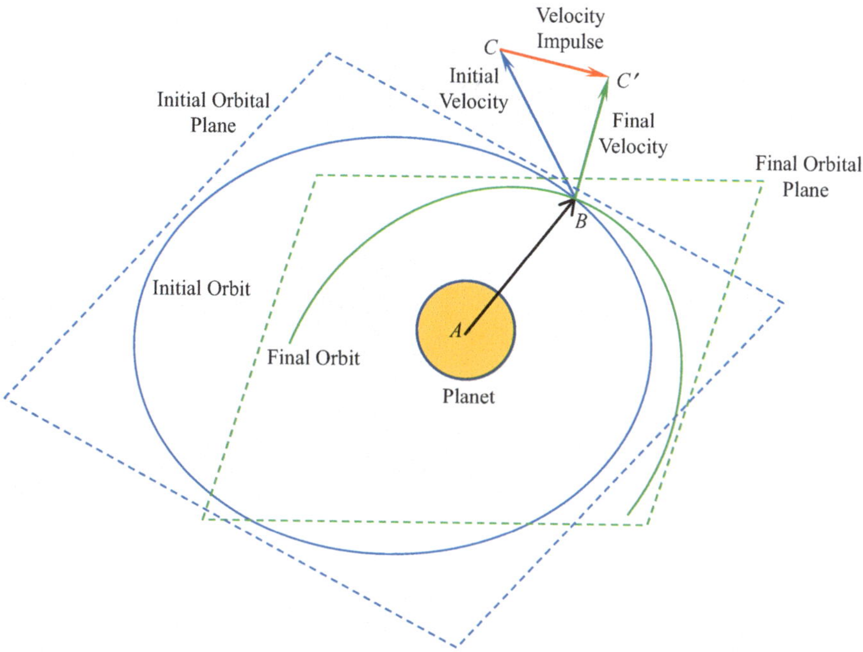

Fig. 6.4 Impulsive orbital maneuver carried out at radius vector *AB*, by applying a velocity impulse *CC′*, causing a change in the shape, size, and plane of an orbit

Space-1 of NASA, have utilized low-thrust orbital transfers to rendezvous missions with asteroids and comets at a much lower mission cost than what would have been required by a conventional chemical rocket.

Other that choosing a rocket with the largest specific impulse, assistance can be taken from the gravity of other orbiting bodies, the atmosphere of a planet, or even the solar radiation.

6.3.2 *Gravity-Assist Maneuver*

A practical way of designing a minimum propellant, interplanetary orbit is by making use of the gravity of intermediate planets for changing the spacecraft's velocity vector. This concept, called a *gravity-assist maneuver*, involves making a close hyperbolic pass of an intermediate planet such that upon emerging on the other side there is a net change in the velocity vector relative to the sun. This would result in a change of the spacecraft's orbit around the sun without expending any propellant. Gravity assist is like a slingshot effect when a planet's revolution around the sun is used to impart a velocity to the spacecraft. Figure 6.5 depicts a gravity-assist maneuver. The planet's gravity essentially applies a velocity impulse

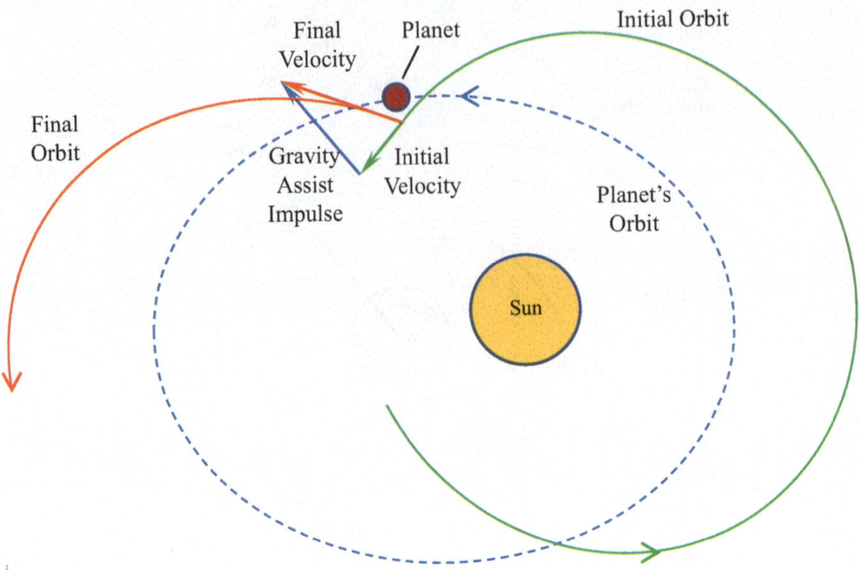

Fig. 6.5 Gravity-assist maneuver. A spacecraft making an interplanetary voyage around the sun uses the gravity of an intermediate planet to change its velocity vector (speed and direction) without spending any propellant in the process

to the spacecraft as it makes a close hyperbolic pass around it. This results in an almost instantaneous change in both speed and direction, producing a new orbit of the spacecraft around the sun. Many spacecraft have utilized multiple gravity-assist maneuvers to significantly reduce the overall cost of their mission. Examples include the *Voyager-I*, *Voyager-II*, *Cassini*, *Galileo*, and *New Horizons* missions to the outer planets. Of these, the two *Voyagers* have experienced such a boost in their respective orbital energies due to repeated gravity-assist maneuvers that they have now escaped from the solar gravity on hyperbolic trajectories relative to the sun, and are presently in the interstellar space.

6.3.3 Aero-Assisted Orbital Transfer

When a planetary atmosphere is employed to make changes in a spacecraft's orbit, the concept is referred to as *aero-assisted orbital transfer* (AOT). There is an atmosphere clinging to most planets which becomes rapidly thinner as the altitude increases. Thus at an orbital radius close to the planet, one may expect to find a small atmospheric density which vanishes at larger radii. Because the orbital speeds are quite high, and the aerodynamic forces are proportional to the square of the relative speed (see Chap. 2), even a very small density can produce aerodynamic forces (lift and drag) large enough for carrying out orbital maneuvers. Consider

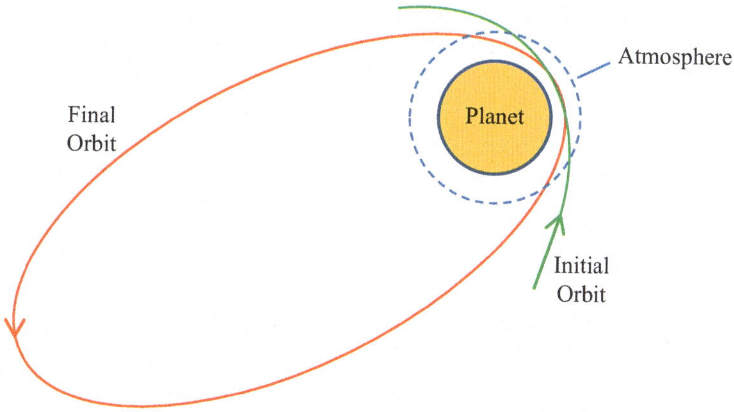

Fig. 6.6 Aero-assisted orbital transfer around a planet. The highest speed occurs at the smallest radius (periapsis) where the largest aerodynamic forces are experienced, thereby causing an almost impulsive change in the orbit

a spacecraft making a close pass of the planet such that it momentarily dips into the sensible atmosphere. Since the speed at the lowest point in the orbit around the planet (periapsis) is the highest, the atmospheric forces experienced by the spacecraft at the periapsis are the largest. As most spacecraft are equipped with large solar panels, they can modify the aerodynamic forces by appropriately orienting the panels relative to the flight path. Depending upon the panel configuration, the angle-of-attack and the bank angle, the aerodynamic lift and drag can be used to decrease the speed while also changing the flight direction. This results in a new orbit of the spacecraft, as it emerges from the atmospheric influence. Because the most significant change in the orbit takes place only at the periapsis, it is as if a velocity impulse has been applied to the spacecraft at that point. Figure 6.6 illustrates the AOT concept. The AOT can be utilized to either change the orbit from open (hyberbolic) to closed (elliptical), or from an elliptical to a circular orbit. The former concept is called *aerocapture*, while the latter is referred to as *aerobraking*. The same maneuvers would require an enormous propellant expenditure, which would increase the cost of the mission several times. The *Magellan* mission to Venus was the first spacecraft to experimentally explore aerobraking, while the *Mars Global Surveyor* (MGS) mission had aerobraking as a part of its main mission, which involved making multiple atmospheric passes of Mars carried out over several months. However, since the orbital energy is lost in each aerobraking pass which causes a reduction in the periapsis radius, a small rocket boost must be periodically applied at the largest radius (apoapsis) point to prevent the spacecraft from crashing into the planet as its orbit is being circularized. Aerocapture is much riskier than aerobraking, because it requires a spacecraft to descend into the denser portion of atmosphere, where the large heating rate due to atmospheric drag can cause a catastrophic structural failure. Therefore, aerocapture requires that the spacecraft must be equipped with a heat shield (or at least an ablative coating), which is an exterior shell of special heat-resistant materials for absorbing the high temperatures without transmitting heat to the spacecraft's components. Once the spacecraft is successfully captured into the elliptical orbit, the heat shield can be jettisoned. The only spacecraft which have experimented with aerocapture so far were the Soviet *Zond*-6 and *Zond*-7 unmanned capsules, which used aerocapture around the earth to shed excess velocity before re-entry while returning from the moon. *Zond*-6 was destroyed in the process, while *Zond*-7 made a safe landing.

6.4 Orbital Perturbations

There are always some small forces acting on a spacecraft apart from the spherical gravity of the primary body it is orbiting. These small secondary forces act as disturbances which prevent the spacecraft from following a fixed orbit. An example of the perturbing force is the atmospheric drag, which can arise due to a very thin planetary atmosphere extending into the space (see Appendix). As we know from the previous chapters, drag always acts in the opposite direction to the motion of any object, and hence tends to slow it down. When a small drag of a thin atmosphere is acting continuously on a satellite orbiting close to a planet, the satellite continuously loses its energy with time (albeit very slowly) and has to be boosted up by periodic firings of its rocket engine. If such boosts are not provided, the satellite will eventually lose its altitude so much that it re-enters the dense lower part of the planetary atmosphere, and burns up. However, every energy boost by the rocket engine depletes the propellant store, which gets completely exhausted after a period of time. Then there is no way of maintaining the spacecraft in its orbit. Hence, the life of a satellite in an orbit is determined solely by the amount of propellant it has on board.

Perturbing forces other than the atmospheric drag include those of a non-spherical gravity. As we saw in the previous section, when the spacecraft is orbiting a spherical body, the gravity always acts toward the planet's center, resulting in a conic section orbit of a constant shape, size, and plane. If either the planet is not exactly spherical in shape, or if there is a gravitational pull from another body, the net gravity acting on the spacecraft is not always directed toward the planetary center. Hence the spacecraft departs from a constant conic section orbit. Such a perturbed orbit, being caused solely by gravitational forces, always maintains a constant average energy per orbit and hence does not experience a change in the orbital shape and size. For this reason, the non-spherical gravity is said to be a *conservative perturbation*. The non-spherical gravitational perturbation is such that it causes a precession and a rotation of the orbital plane as explained next.

In order to understand the effects of non-spherical gravity, let us begin with the case when the primary body is not exactly spherical in shape. Such a body produces a non-uniform gravitational field in all directions. An example of such a body is an *oblate* planet like the earth, which has a larger radius at the equator compared to that at the poles. Being bulging in the middle, such a planet exerts a torque on the

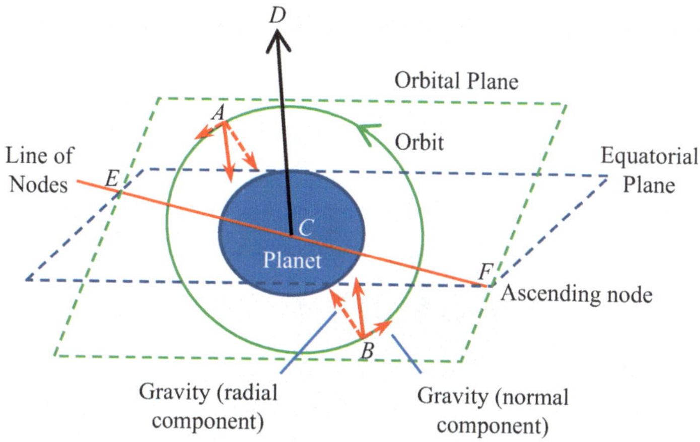

Fig. 6.7 Out-of-plane gravity component applying a pure couple on the orbital plane of a satellite in circular orbit due to an oblate planet

orbital plane of any spacecraft which is not exactly in an equatorial orbit. Figure 6.7 shows a spacecraft orbiting an oblate planet in a circular orbit inclined from the equator. At any point away from the equator, say the point A in Fig. 6.7, there is an off-center gravitational tug on the spacecraft, which results in a gravity component normal to the plane of the orbit. Of course, the orbit being symmetrical in shape, there would be another (diametrically opposed) point (B) in the orbit where the out-of-plane (normal) gravity component is equal in magnitude, but opposite in direction to that at point A. The radial components at the two points are acting in a common line, thus cancel each other. Since the gravity components normal to the orbit at the points A and B form a pure couple, a torque is applied on the orbital plane normal to the orbital axis (depicted in Fig. 6.7 by the arrow CD). Since this happens at all the points which are not in the equatorial plane, the spacecraft's orbit would continuously experience a perturbing torque normal to its axis, causing it to precess like a spinning top. Consequently, the orbital spin axis describes the dashed cone shown in Fig. 6.8, and the plane of the orbit is continuously changing with time. As a result, the *line of nodes*—defined as the intersection of the orbital plane with the equatorial plane, and depicted in Fig. 6.7 by the line EF—also rotates with time in the sense of the orbit as shown in Fig. 6.8. Note that the point F in Fig. 6.7 is the *ascending node* of the orbit, which shows the direction from the planet where the orbit crosses the equatorial plane from the south to the north. Similarly, the point E is the *descending node*. For an inclination angle less than 90°, the rotation of the line of nodes is from the west to the east, which is opposite to the direction of the earth's rotation. Therefore, such an orbital precession is said to produce a *regression of nodes*. There is no change in the angle of inclination and the orbital shape and size.

If the spacecraft orbit in Fig. 6.7 is elliptical rather than circular, there would also be an in-plane rotation produced by the oblate shape of the planet. To understand this, we note that the largest gravitational tug occurs at the smallest radius (periapsis)

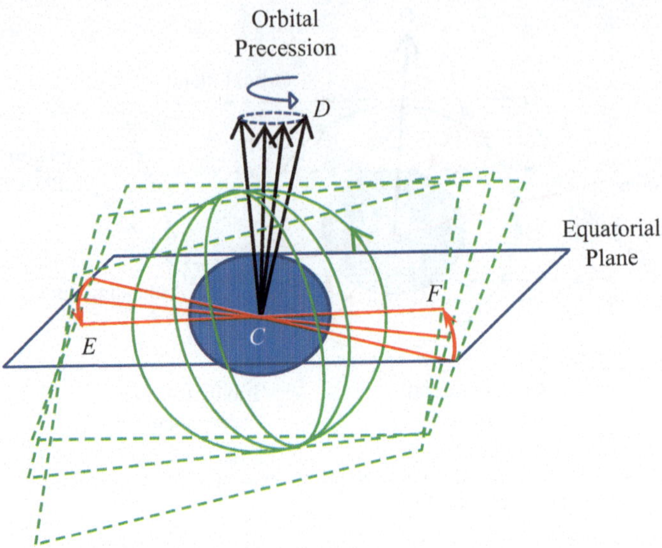

Fig. 6.8 Precession of the orbital plane of a satellite in a circular orbit due to an oblate planet producing a regression of nodes

of the ellipse. But the opposite point in the elliptical orbit is at the largest radius (apoapsis), which experiences the smallest gravity. Hence, there is now a difference in the magnitudes of the perturbing normal force component at the two radially opposed points. Extending this to a general pair of radially opposed points, say *A* and *B*, we can see that the normal components do not form a pure couple, which was the case in the circular orbit of Fig. 6.7. Hence, in addition to the torque per orbit normal to the spin axis that causes precession (Fig. 6.8), there is now another torque continuously acting in the plane of the orbit, which causes the entire orbital plane to rotate about its own axis. The in-plane torque component averages to zero over a complete orbit, hence in each orbit there is a rotation of the line joining the periapsis and the apoapsis by a constant angle (called the *line of apsides*, depicted as the line *GH* in Fig. 6.9. This effect is referred to as the *rotation of apsides*. Therefore, the location of the periapsis (the point *G*) in the elliptical orbit keeps on changing relative to the equatorial plane. Figure 6.9 shows the regression of nodes and the rotation of apsides for an elliptical inclined orbit around an oblate planet.

The effects of third-body gravity are similar to that of an oblate planet, because they are also produced by a non-spherical gravity field. The difference is that the perturbing force is now a function of the angular position of the perturbing body relative to the spacecraft, which of course, keeps on changing with time. Since the third body is usually very far away from both the primary body and the spacecraft, its distance can be taken to be essentially a constant at its average value. Therefore,

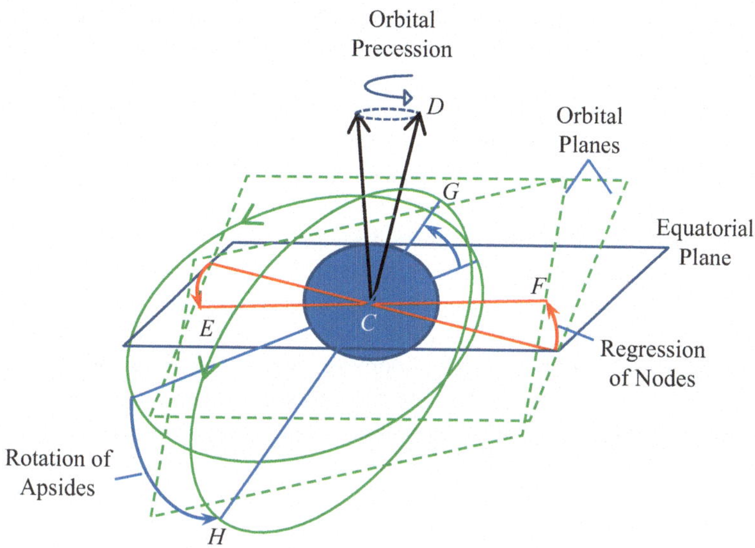

Fig. 6.9 Rotation of apsides and regression of nodes of a satellite in an elliptical orbit due to an oblate planet

the rotation of apsides and the regression of nodes happen at rates which depend upon the average distance and the angular deviation of the spacecraft from the perturbing body.

6.5 Geosynchronous and Sun-Synchronous Orbits

If a satellite is in an equatorial circular orbit of a radius such that the orbital rate equals the planet's rate of rotation about its polar axis, then the satellite would appear stationary to an observer on the planet. This is the concept behind the *geosynchronous orbit*. For earth, a geosynchronous satellite must have an orbital radius of about 42,164 km in order to produce an orbital rate exactly equaling the earth's rotational rate on its axis. The geosynchronous satellites are used as telecommunications platforms to relay radio and television signals from one point on the earth to another.

The earth takes exactly 23 h, 56 min, and 4.09 s to complete one rotation on its axis against the background of distant stars. This is called the *sidereal period*, which is different from the 24 h period required to complete a rotation relative to the sun. The difference in the two periods is due to the fact that the earth also revolves around the sun in the same direction as it rotates (from the west to the east). Hence, the earth is also traversing an angle of about 360/365° per day (called the *diurnal rate*) relative to the sun, which must be added to the total angle covered while calculating the sidereal rotational rate.

Imagine a satellite orbit of constant radius around the earth, which is inclined to the equator such that the regression of nodes caused by the earth's oblateness is negative in sign, and equal in magnitude to the diurnal rate (360/365° per day, east to west) with which the sun appears to move relative to the earth. This would require the satellite to have an orbital inclination greater than 90°, which makes it orbit the earth in a sense opposite to earth's rotation (i.e., the satellite appears to move from east to west relative to the earth). Such a satellite will maintain a *constant* orbital inclination *relative to the sun*, and is called a *sun-synchronous satellite*. This is very important from photographing purposes because all the points covered by a sun-synchronous satellite are illuminated at a fixed solar angle, which makes the identification of objects on the ground much easier. If an object is photographed at a different illumination angle from above, it appears to be completely different because of the change in the length and direction of the shadow it casts on the ground. Therefore, a sun-synchronous satellite can be used to compare photographs taken of a given object (such as a building, or a vehicle) at different times in order to see whether it has changed due to any reason. Such changes would not be detected if the illumination angle was not a constant.

6.6 Atmospheric Entry Vehicles

Many spacecraft (especially the manned ones) have to return to a planetary surface as a part of their mission. For example, a spacecraft sent to land on the planet Mars must enter the Martian atmosphere and slow down sufficiently to make a safe landing on its surface. The astronauts returning from a space station to the earth must similarly undergo a flight termed as the *re-entry*. Because a spacecraft is initially traveling at a speed of several kilometers per second, the problem of aerodynamic heating is paramount while decelerating in the atmosphere. Such a heat is generated both by the skin friction, as well as by strong shock waves that compress the air in the front of the vehicle (much like a piston in an automobile engine), thereby raising its temperature. For example, the *Apollo* entry capsule returning the astronauts from the Moon faced a peak temperature that exceeded even that found at the surface of the Sun! A safe entry flight through the atmosphere therefore requires that the excessive heat produced in the process must be dissipated without destroying the structure and the payload (often, human beings).

6.6.1 Hypersonic Aerothermal Load

The vehicle experiences different flow regimes as it enters deeper into the atmosphere. Near the outer edge of the atmosphere, the vehicle encounters rarefied atmospheric gases with highly variable and uncertain chemical properties. Since the gas particles are separated by large distances, they do not interact with each other significantly after contacting the vehicle. Hence, the vehicle experiences a *free-molecular flow* near the outer portions of the atmosphere, which can be understood by considering how individual particles interact with the solid surfaces of the vehicle.

As the vehicle comes closer to the planetary surface, it sees an almost continuous gaseous medium due to its higher density and nearly constant chemical properties. Now the flow experienced by the vehicle is that of a *continuum*, where individual particles need not be considered and only the bulk properties of the medium (such as density, viscosity, etc.) are important. In the continuum flow regime, the higher density and viscosity of the fluid give rise to a much higher pressure, skin friction, and heat transfer than that possible in the free-molecular regime. Between the two extremes of free-molecular and continuum flow, we have a *transition flow* regime, where the flow is governed both by individual particle and bulk medium properties. the flow parameter governing rarefied flows is the *Knudsen number*, defined as the ratio of an average separation of the molecules (the *mean free path*) in the freestream to a characteristic length of the vehicle. The Knudsen number for a typical entry trajectory can range from greater than 100 (free-molecular flow) to less than 1/100 (continuum flow).

The largest forces and rates of heat transfer due to the atmosphere are encountered in the continuum regime. The combination of the two effects is termed the *aerothermal load*. The rate of heat transferred to the vehicle depends crucially upon the geometry of the vehicle, the material properties of its surface, as well as the speed and the steepness of its trajectory through the atmosphere. A trajectory that rapidly traverses the continuum regime would thus have a much higher aerothermal load than the one which does so much more slowly.

The peak Mach number in the continuum flow around an entry vehicle can range between 15 and 35. At such high Mach numbers, the aerodynamics is completely different from that seen around an airplane, and includes the high-temperature gas dynamics, where the air ceases to behave as a perfect gas and undergoes thermal and chemical transformations. The following features dictate the distribution of pressure and temperature on a vehicle in a hypersonic flight:

(a) All the changes in the flow around the solid body are confined in a narrow region close to the surface called the *thin shock layer*. The shock wave envelopes the body very closely at the hypersonic Mach numbers of entry. Due to this fact, the flow variations are much more tightly packed around the body, and result in much higher temperatures when compared to those at smaller Mach numbers. Figure 6.10 schematically depicts the hypersonic flow encountered in the continuum regime on a vehicle shaped like a cone, but with a rounded nose. The complex flow patterns caused by shock-wave/boundary-layer interaction (Chap. 2), reflected expansion waves, and secondary shock waves—all of which are present in the shock layer—are not shown here.

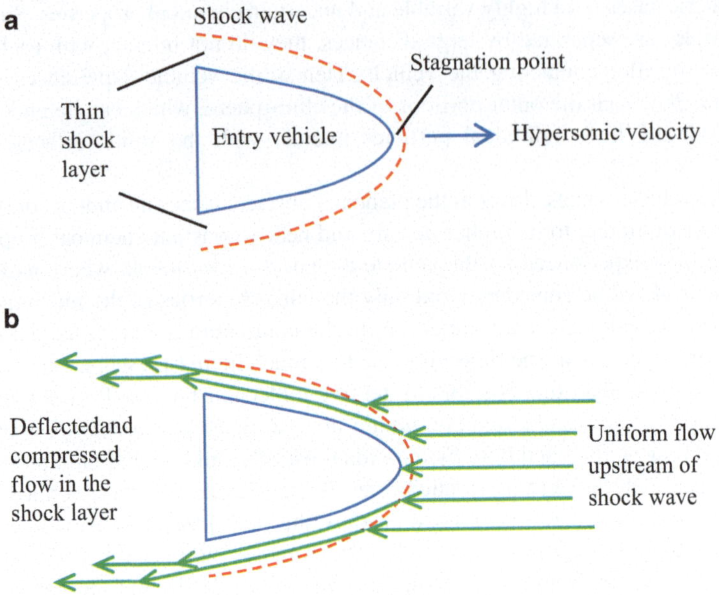

Fig. 6.10 (**a**) Hypersonic airflow around an entry vehicle passes through a thin shock layer formed between a strong, bow-shaped shock wave which is tightly wrapped around the body. (**b**) The flow is uniform outside the shock layer, and all the pressure and temperature variations take place inside the shock layer. The combined effects of compression due to the shock wave and skin friction due to the body generate heat in the shock layer. The rate of heat transfer is the maximum at the stagnation point

(b) The rate of convective heat transfer is the maximum at the stagnation point (see Chap. 2), and varies inversely with the square-root of the local radius of curvature at that point. This means that a sharp stagnation point—that is, nearly zero radius—sees an almost infinite rate of heat transfer at hypersonic velocities. For this reason, all entry vehicles must present as large a radius as possible at the stagnation point. The sharp nosed objects which are optimized for supersonic flight are thus entirely unsuitable at hypersonic speeds. The *Space Shuttle* of NASA had snub-nosed fuselage and a rounded leading edge of its wing, and entered the atmosphere at a high angle-of-attack which placed the stagnation point at the flat part of the wing fuselage junction. This resulted in a nearly infinite radius at the stagnation point, and consequently the minimum possible heating rate. For the same reason, the bottle-shaped entry capsules of the *Apollo* and *Soyuz* type orient themselves to present the flat part of their bottom as the stagnation point.

(c) The rate of heat transfer experienced by the vehicle at any point in its trajectory varies directly with the cube of its speed at that point. This happens due to the increased strength of the normal shock wave at the stagnation point. For example, a spacecraft entering the earth's atmosphere from a low-orbit has an entry speed of about 8 km/s, while that of a lunar return vehicle is closer to 11 km/s. A spacecraft returning from Mars could be expected to approach the earth at nearly 13 km/s. This implies that a lunar return vehicle would experience more than 2.5 times the peak heating rate of a spacecraft re-entering from a low-earth orbit, and a Mars-return spacecraft would have an entry heating rate more than four times that value.

(d) A steep entry trajectory has a much higher rate of heat transfer, everything else remaining the same, than a shallow entry trajectory. However, if the flight path angle at entry is too shallow there is the possibility of the vehicle skimming the dense portion of the atmosphere, and exiting the atmosphere altogether on a *skip trajectory* much like a stone skipping over the surface of a pond. This could be dangerous in a manned mission, because an unplanned skip trajectory greatly increases the time of flight for a space capsule with a rather limited oxygen supply. The correct angle of entry should be chosen as the one which results in the minimum heating rate as well as does not lead to a skip entry.

(e) The steeper the descent into the atmosphere, the larger is the deceleration due to the drag. Thus the load experienced by an entry vehicle depends upon its flight path angle. A shallower entry angle would thus produce both a smaller heating rate and a smaller deceleration.

6.6.2 Ballistic and Lifting Trajectories

It is extremely important for the safety of an entry mission to follow a trajectory which minimizes the aerothermal load on the vehicle at any given point, and also results in a sufficient reduction in the final velocity for a safe landing. If the vehicle has a lifting capability like the *Space Shuttle*), then its flight path angle can be controlled by generating a normal load factor. The resulting *lifting-entry* trajectory is much shallower than a typical non-lifting (ballistic) entry trajectory, and delays the build-up of the aerothermal loads by keeping above the denser portion of the atmosphere, until the speed has decayed sufficiently. This might allow the vehicle to travel a greater horizontal distance before landing. For example, the *Space Shuttle's* lifting-entry trajectory took it over nearly half of the earth circumference.

Figure 6.11 compares the ballistic and lifting entry trajectories for a vehicle returning to the earth from a low orbit. Both the trajectories originate at the same point in the orbit at the same initial speed. The center of mass of the vehicle at a common distance (radius) from the earth's center A is denoted as the point B, whereas the end of the velocity vector is marked C. The local horizon is the plane normal to the local radius vector, and the flight path angle is the angle $\angle CBD$ made by the velocity vector with the local horizon. Hence, $\angle CBD$ is the *complementary* (i.e., the angle subtracted from 90°) of the angle $\angle ABC$ between the local radius vector and the velocity vector. The smaller the angle $\angle ABC$, the larger the flight path angle $\angle CBD$, thus steeper the trajectory. Note that the ballistic entry trajectory is the steeper of the two, because it always has a larger flight path angle $\angle CBD$

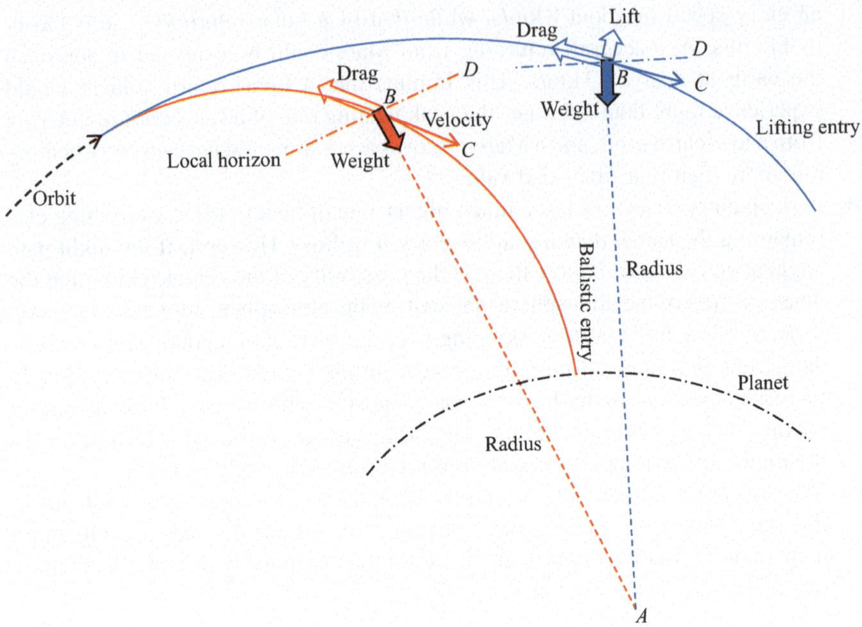

Fig. 6.11 Ballistic and lifting entry trajectories. A ballistic trajectory experiences a higher aerothermal load at a given time after entry, due to its encounter with the dense portion of the atmosphere at a higher speed

at a given radius. Consequently, at a given time after entry, the vehicle following a ballistic trajectory has penetrated deeper into the atmosphere and experiences a larger drag and heat transfer. Comparatively, the lifting trajectory stays above the ballistic one, therefore sees a smaller drag and heat transfer at any given time.

When the vehicle has slowed down sufficiently at the end of its entry trajectory, retardation devices such as parachutes and retro-rockets can be deployed to decrease its speed before the touchdown. If the vehicle is capable of generating lift, it can carry out additional maneuvers to slow down as well as to make an approach and landing like a glider. The use of retardation devices and maneuvering before the touchdown is referred to as the *terminal-area energy management* (TAEM). Since the vehicle is descending all the time, its potential energy is continuously decreasing, but the kinetic energy is prevented from increasing by the use of retardation devices and energy-sapping maneuvers, such that the total energy prior to the touchdown is kept to a minimum. An example of the TAEM is the use of a steady sideslip (Chap. 4) by glider pilots before the final approach, commonly carried out whenever they find themselves either too high, or too fast for a safe landing.

6.7 Attitude Stability Control of Spacecraft

All spacecraft must maintain a desired orientation in order to perform their basic mission. Most spacecraft carry sensors (cameras, thermal imagers, radiometers, magnetometers, etc.) that must be pointed in specific directions at different times. Furthermore, nearly every spacecraft generates some electrical power from its solar arrays which must have a desired orientation relative to the Sun for the maximum efficiency. Lastly, the payload and the sensor package must be protected from the damaging effects of solar and cosmic radiation in the space where a protecting atmosphere is absent. All of this requires that every spacecraft must have a dedicated stability and control system for rotating and maintaining it in desired orientations in the presence of small torque perturbations. Such a control system is referred to as an *attitude stability and control system*. The attitude control system is required for either maintaining a fixed orientation, or to match the changing orientation of another body, in the presence of the disturbing torques.

6.7.1 *Spin Stabilization*

The space environment is so quiet that it often does not produce any torque of a significant magnitude. This is to be contrasted with the atmospheric flight, where a slightest change in an aircraft's attitude relative to the flight direction can immediately produce a large torque for providing stability and control. Consequently, one has to devise mechanisms for applying a torque to the spacecraft. As will be discussed later, such mechanisms can either utilize the space environment, such as those producing the *gravity-gradient*, *geomagnetic*, and *solar radiation pressure* torques, or the actuators on board the spacecraft, such as *rocket thrusters* and *rotors*. An advantage of the nearly torque-free space environment is that the angular momentum of the spacecraft is almost conserved. Recall from Chap. 1 that Newton's laws of motion dictate a rigid body must have an external torque applied to it in order to change its angular momentum. The torque-free rotation of a rigid spacecraft thus leads to a constant angular momentum vector, with a fixed magnitude and direction. An example of such a rotation is the equilibrium condition of a single-axis rotation at a constant angular velocity, in which the spacecraft continues to spin forever about a given axis passing through its center of mass without any expenditure of energy.

However, neither the spacecraft is exactly a rigid body, nor the space environment is exactly torque-free. Therefore, neither a constant angular moment, nor a constant energy of rotation can be maintained. Every spacecraft has some liquid rocket propellants in its tanks, which dissipate the energy of rotation by a sloshing motion every time there is a change in the angular velocity. A similar dissipation mechanism is provided by the structural flexibility of the spacecraft's appendages. The energy dissipation mechanism due to the spacecraft being only a semi-rigid (rather than rigid) body involves a relative motion of its parts, and is similar to friction.

The time rate of energy dissipation is proportional to the rate of change of the angular velocity. There would be no dissipation in the rotational energy if the spacecraft continues to spin about a fixed axis at a constant rate. However, such an equilibrium state cannot be reached until the energy has reached the minimum possible value. Thus, the semi-rigid mechanism of energy dissipation continues to act until the state of the smallest energy is reached.

For a given angular momentum, the minimum energy state corresponds to the rotation about the axis with the largest moment of inertia. Suppose the spacecraft is shaped like a pencil, initially spinning about the axis along its length (longitudinal axis). The initial state corresponds to the largest possible energy for the given angular momentum, because it takes place about the axis with the smallest possible moment of inertia (the *minor axis*). The smallest energy state would be a rotation about an axis perpendicular to its length (transverse axis) passing through the center of mass, since its has the largest moment of inertia (hence called the *major axis*). If there are no external torques, the initial rotational state would be ultimately transformed into a spin about the transverse axis, which will be a slower rate because the angular momentum (i.e., the product of the moment of inertia and the angular speed) must be conserved.

The discussion given here shows that in a torque-free case, it is possible to achieve an equilibrium state by spinning the spacecraft about its major axis. Any small perturbation applied to the spacecraft in such a state would be naturally dissipated, and the major axis will ultimately return to its original alignment with the angular momentum vector. The angular momentum vector—being fixed in space—thus provides a constant orientation for the spacecraft's major axis. This mechanism of stabilizing a spacecraft's attitude is called *spin stabilization*.

A common application of spin stabilization is for a *nadir pointing* satellite (see Fig. 6.12), where the spin axis is normal to the plane of the orbit, and the spin rate is matched with the constant orbital rate of the specified circular orbit. In this case,

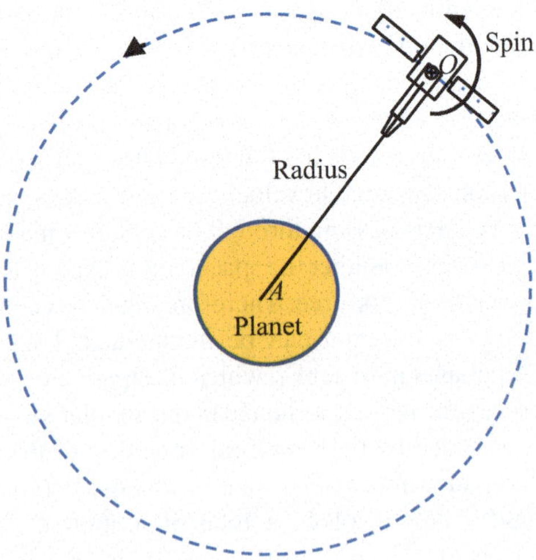

Fig. 6.12 A spin stabilized, nadir pointing satellite in a low circular orbit around a spherical planet. The major axis is normal to the orbital plane, and the orbital rate is synchronized with the spin rate. If the minor axis is aligned with the radius, *OA*, then this orientation also becomes a stable gravity-gradient attitude of the spacecraft

the satellite always points is lowest point toward the center of the planet. A payload, such as a camera or an imaging device, can thus be always aimed at the ground for photography, resource mapping, or imaging.

Many other objects—such as rifle bullets and artillery shells—are also spin stabilized for maintaining a nearly constant orientation for a better targeting accuracy. However, as they are rigid bodies, they do not have an internal energy dissipation mechanism. In addition, the aerodynamic moments do not allow a torque-free motion through the air. Consequently, the stability of the projectile's rotational dynamics is dependent upon the damping produced by the aerodynamic moments.

6.7.2 Gravity-Gradient Stabilization

A spacecraft experiences a tiny torque due to the differences in the gravity at the various points along its length. This torque is termed the *gravity-gradient* torque. While every object experiences such a torque, it is noticeable only in the quiet space environment, where it can be used to stabilize the rotational motion of a spacecraft. The appreciable magnitude of such a torque is produced only when the spacecraft is in a low orbit.

Consider a satellite in a low circular orbit around a spherical planet. If the major axis of the spacecraft is normal to the orbital plane (as in a nadir pointing satellite), while its minor axis points either toward, or directly away from, the planetary center (as shown in Fig. 6.12), then a naturally stable equilibrium state is reached where the gravity-gradient torque opposes any small disturbances. This is a very useful stabilization technique for large spacecraft—such as space stations—which encounter a significant gravity-gradient torque due to their length.

If a spacecraft is not in the stable gravity-gradient orientation described above, it will have to spend energy in fighting the gravity-gradient torque to remain in equilibrium. Such a spacecraft can ran out of the propellants (thereby shortening its life) while struggling to maintain an equilibrium orientation against the gravity-gradient torque.

6.7.3 Attitude Control

In order to change the angular momentum of a spacecraft, there must be a torque applied about its center of mass. There are basically three ways of generating the torques necessary for rotating a spacecraft as desired: (a) rocket thrusters, (b) rotors, and (c) magnetic torquers. Of these, the rocket thrusters are easily understood as suitably placed pairs of rockets which can apply equal and opposite forces to create a net moment about the center of mass. A minimum of three pairs of rockets about three mutually perpendicular axes (pitch, roll, and yaw) are necessary for a complete

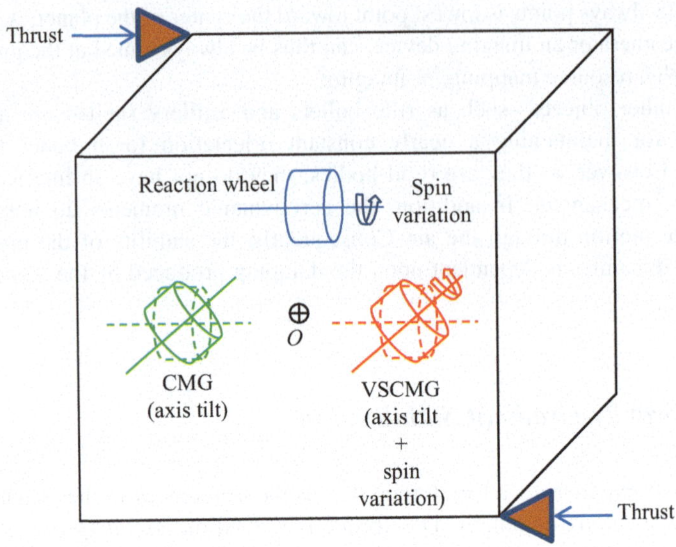

Fig. 6.13 A schematic diagram of a spacecraft equipped with a pair of rocket thrusters, a reaction wheel, a control moment gyro (CMG), and a variable-speed control moment gyro (VSCMG). The rocket thrusters apply an external torque about the spacecraft's center of mass, O, whereas the rotors produces an internal torque by the principle of angular momentum conservation. (This is only for illustration purposes, because an actual spacecraft can have many more of such devices.)

control of the spacecraft's attitude, as shown in Fig. 6.13. However, the rocket thrusters deplete the propellant store every time they are fired thereby decreasing the spacecraft's life in orbit, therefore are not to be used for routine attitude stabilization tasks which require torques to be produced repeatedly upon demand.

Rotors are much more efficient than rocket thrusters as they only require the electrical power to run, which can be generated by a spacecraft's solar arrays (or in some cases, radioactive thermal generators). Their repeated use for attitude control does not affect the life of the spacecraft. The rotors generate an internal torque about the spacecraft's center of mass. The creation of an internal torque by a rotor is possible because the total angular momentum of the spacecraft plus the rotor must be a constant, as there are no external torques. A change in the rotor's angular momentum thus causes an equal and opposite change in the spacecraft's angular momentum. The rotors used for spacecraft attitude control can be divided into three categories:

1. *Reaction wheel* with a fixed spin axis relative to the spacecraft, but having a variable speed rotor. Such a rotor generates a control torque about its spin axis.
2. *Control moment gyro* (CMG) with a variable spin axis relative to the spacecraft, but having a constant speed rotor. Such a rotor generates a control torque normal to its spin axis.

3. *Variable-speed control moment gyro* (VSCMG) with a variable spin axis relative to the spacecraft, and also having a variable speed rotor. Such a rotor generates a control torque in a general direction about the spacecraft's center of mass.

The three types of rotors are schematically shown in Fig. 6.13.

For the complete controllability of the spacecraft in pitch, roll, and yaw, a minimum of three reaction wheels are required (one per axis). In contrast, only two CMGs are necessary for a multi-axial torque. The main advantage of a VSCMG is that a single rotor can theoretically apply a torque about all the three axes. However, practical considerations require a redundant set of rotors to account for the situation in which some rotors fail to produce the necessary torque. A pyramid configuration of five to six rotors is usually employed, with the spacecraft's center of mass as its centroid.

Since the rotors have a small mass relative to that of the spacecraft, they must be spinning at very high rates in order to produce an appreciable torque. Otherwise, the torque generated by them would be too small for maneuvering the spacecraft in a reasonable time-frame. Recall that the angular momentum of a rotor is the product of its spin rate and its moment of inertia. Since the latter is quite small, the former must be rather large for a given rate of change of angular momentum. However, there is a structural limit for the maximum speed of each rotor. A reaction wheel becomes *saturated* when its speed reaches a limiting value; thus it cannot increasing its speed any further to produce a desired torque. The wheel must therefore be spun down to zero speed in order to restore its utility. However, if this were to be done without canceling its torque by another device then an unwanted torque would be applied to the spacecraft. Hence, either a pair of rocket thrusters, or another set of rotors should be available to de-saturate the reaction wheels.

A CMG rotor having a constant spin rate, is free from the saturation problem, but has its own practical limits due to a limitation on the tilt angle of its axis.

Similarly, a VSCMG rotor can encounter a difficulty in generating the desired multi-axial torque when its rotor attains a particular orientation, in which the torque is actually produced only about one axis of the spacecraft. This is called the *gimbal-lock* phenomenon.

Lastly, a *magnetic torquer* is a device for utilizing the magnetic field of the primary body to produce a torque about a satellite in a low orbit. This device consists of an electromagnetic coil powered by the spacecraft's own electrical system. As the spacecraft crosses the magnetic field, a torque is produced whose strength depends upon the inclination of the orbital plane with respect to the equator, and the orientation of the spacecraft relative to the flight direction. While the torque produced by a magnetic torquer may not be sufficient to maneuver a spacecraft, it can be used in conjunction with the spacecraft's rotors to provide an active damping of any oscillations in its attitude dynamics. Thus, the dynamic stability in a stable equilibrium state can be achieved with the help of a magnetic torquer.

Chapter 7
Rocket Flight

7.1 Introduction

Rockets are necessary to enable space flight, because any other form of propulsion cannot impart the necessary amount of velocity to put a spacecraft into orbit. Another important reason that rockets are indispensable for space flight is that they can provide thrust without requiring the presence of an atmosphere. Chemical rockets are typically employed to produce both a large thrust and a reasonable specific impulse by using a chemical reaction in a combustion chamber that generates an enormous amount of heat. This heat is then used to accelerate the resulting gases in a specially designed nozzle, which converts the thermal energy into kinetic energy of the exhaust. By Newton's third law of motion, the exhaust gases apply an equal and opposite reaction on the rocket (the thrust) which causes it to accelerate in a direction opposite to that of the nozzle exhaust. A chemical reaction requires at least two substances, called the *fuel* and the *oxidizer*. The combined mixture of the fuel and the oxidizer is called the propellant, which results in a combustion. The propellant can either be a solid mixture which is ignited by a heat source, or two liquids which must be stored separately in tanks before being pumped into the combustion chamber and ignited. For example, a crude firecracker rocket has a mixture of powdered charcoal and sulfur as the fuel, and potassium nitrate as the oxidizer. A more sophisticated solid propellant mixture is fine aluminum powder (fuel), and ammonium perchlorate (oxidizer). Examples of liquid propellants are kerosene (fuel) plus liquid oxygen (oxidizer), hydrazine (fuel) plus nitrogen tetra-oxide (oxidizer), and liquid hydrogen (fuel) plus liquid oxygen (oxidizer). Solid rocket propellants are easier to store than liquid propellants, and do not require a sophisticated pumping mechanism. However, once ignited, it is difficult to stop the solid rocket combustion, while the combustion of liquid propellants can be controlled by manipulating the volume of the propellant pumped into the engine (called *throttling*). Thus liquid propellants are favored over solid propellants in space flight missions which require intermittent (or impulsive)

A. Tewari, *Basic Flight Mechanics*, DOI 10.1007/978-3-319-30022-1_7

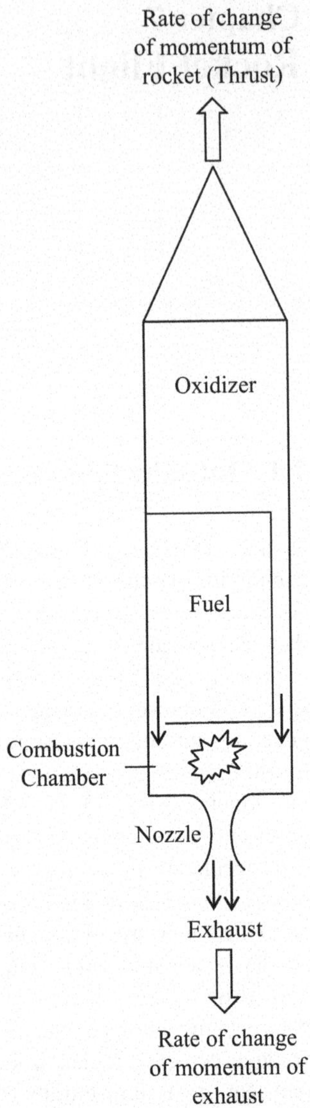

Fig. 7.1 A chemical rocket engine generates thrust by burning its propellant in a combustion chamber, and exhausting the resulting hot gases in a nozzle at a high velocity, thereby propelling the rocket in the opposite direction by Newton's third law

firings. The chemical rocket principle is depicted in Fig. 7.1. Variations of the rocket principle are the acceleration of charged gases in an electromagnetic nozzle (plasma rocket), and nuclear propulsion wherein the heat is produced by a nuclear reaction rather than by a chemical combustion. However, these alternative forms of propulsion can currently produce only a very small thrust and hence cannot be used to launch a spacecraft into orbit.

7.2 Launch Vehicle Staging

A rocket used to launch a spacecraft into an orbit is called a *launch vehicle*. Since a large velocity change must be imparted in order to launch some thing into an orbit, such rockets must carry an enormous amount of propellants. For the maximum efficiency, such rockets must jettison a part of the structure used to carry the propellant as soon as its portion of propellant has been consumed. As soon as the propellant of a stage is completely exhausted, it is jettisoned from the rocket and the lighter resultant structure is propelled much faster by the remaining stages. This principle is called *staging*, and results in a much larger overall specific impulse when compared to a single-stage rocket in which there is no mechanism available to jettison the stages. The concept of staging is schematically depicted in Fig. 7.2. Most launch vehicles have three to four stages, because having more than four stages does not give any appreciable advantage. Each stage has its own propellant (which may be different from the other stages) and its own engine. The second stage forms the "payload" of the first stage, the third stage is the payload of the second stage, and so on. The last stage has the spacecraft to be launched as its payload. Often, the larger launch vehicles also have a number of rockets attached externally to the vehicle. These are called *strap-on boosters* and are burned in parallel with the first stage. While the strap-on boosters are in operation, the resulting stage is referred to as the zeroth stage. For example, the erstwhile *Space Shuttle* of NASA had two large solid rockets as strap-on boosters, which were burned along with the main rocket engine whose liquid propellant was externally stored in a large fuel tank.

7.3 Gravity-Turn Trajectory

For launching a spacecraft into orbit, it is necessary to have a continuously variable flight direction, from exactly vertical at the time of launch to an specified angle when the vehicle has reached the orbital radius. For example, when launching to a circular orbit, the final velocity vector must be perpendicular to the radius vector, which implies that the vehicle must be traveling in a horizontal direction at the time of releasing the spacecraft. As the attitude control system must always keep the vehicle aligned with the velocity vector (see the discussion above), there is no possibility of changing the flight direction by generating a normal force (lift or sideforce). However, the gravity comes to our rescue in this case. The gravity applies a force component normal to the trajectory as soon as the flight direction veers from exactly vertical (i.e., along the radius vector). The magnitude of the normal gravity component increases as the angle between the flight direction and the local radius vector increases. Hence, a natural feedback mechanism is provided by gravity to automatically change the flight direction without requiring any thrust inputs! The naturally curving ballistic trajectory under the sole influence of gravity is called a *gravity-turn trajectory*. Figure 7.3 depicts the launch of a rocket from

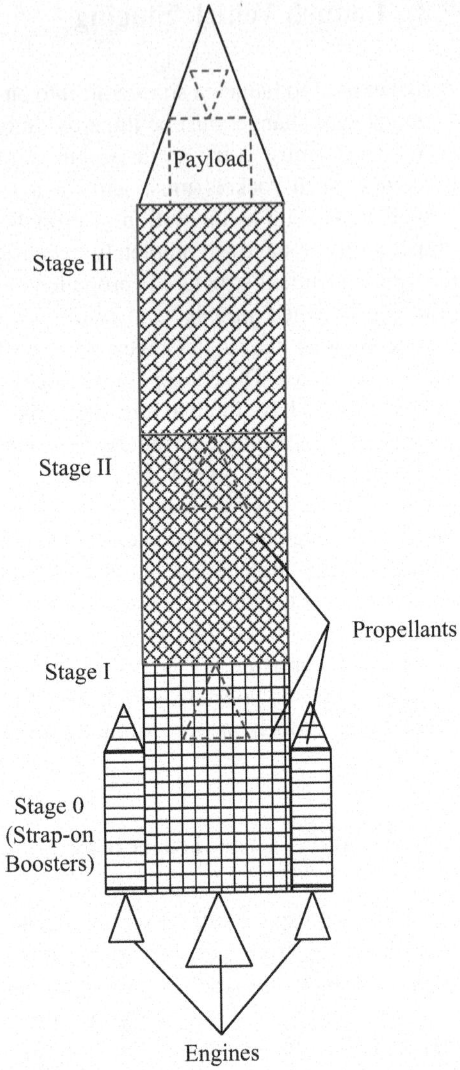

Fig. 7.2 Schematic diagram of a launch vehicle showing a multi-stage rocket, with each stage equipped with its own engine and having its own propellant. *Stage 0* refers to the entire rocket; *stage I* is the part of the rocket remaining after the strap-on boosters have been jettisoned; *stage II* is the remainder after stage I has been jettisoned, and so on

the surface of a planet to an orbit via a gravity-turn trajectory. Immediately after vertical launch from the planetary surface, the rocket is slightly "nudged" in the required direction, and then the gravity takes over, causing the trajectory to turn by its component normal to the flight path (the arrow marked *BD* in Fig. 7.3). For the vehicle to accelerate forward along the trajectory, the thrust must be large enough to overcome the sum of the tangential gravity component (the arrow *BE*) and the atmospheric drag. The attitude control system must rotate the vehicle at exactly the same rate at which the flight direction is changing. This is called the *pitch-program*, and is implemented by having a pitch rate feedback by a rate gyro (to be discussed

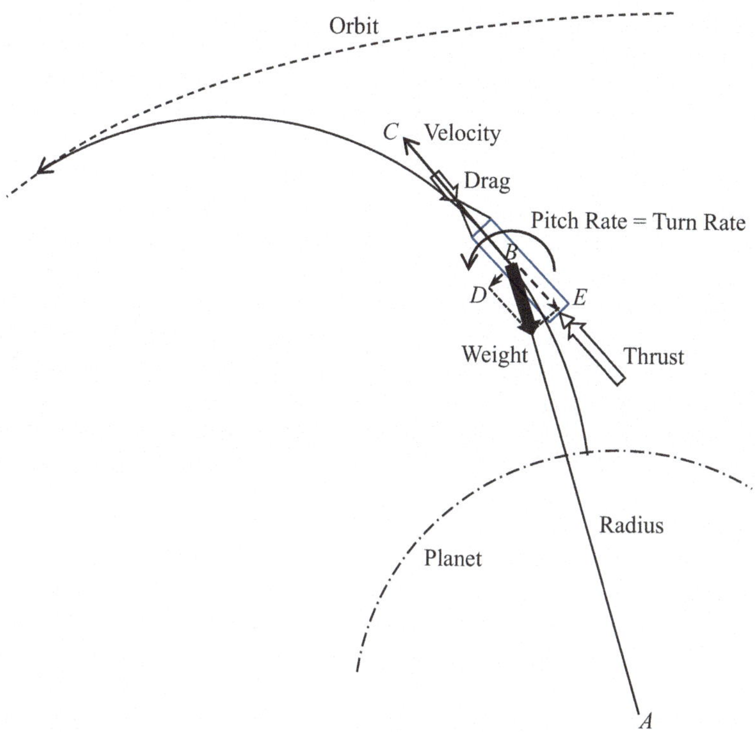

Fig. 7.3 Gravity-turn trajectory for a ballistic rocket launch to a circular orbit. The gravity provides a natural normal force component, which increases with the increasing angle $\angle ABC$ until a horizontal flight direction ($\angle ABC = 90°$) is finally reached at the required orbital radius

later). The pitch rate is then synchronized by the controller with the sensed rate of change of the flight direction, which is provided by the vehicle's guidance system. The timing is such that as soon as all the propellant is exhausted, the vehicle reaches the desired flight velocity and radius. When sophisticated guidance and control systems were unavailable in the past, the task of synchronizing the turn rate with the pitch rate along the gravity-turn trajectory was carried out by a simple clockwork mechanism, such as that used in the German *V-2* rocket of World War II.

7.4 Attitude Stability and Control

Like the aircraft and spacecraft, rockets too require a means of keeping them in equilibrium in the presence of disturbances (stability), as well as the capability to change the equilibrium state as desired (maneuverability). Both stability and maneuverability require that the entire vehicle should be rotated about an arbitrary axis passing through its center of mass. There are two methods of applying forces and moments to a rocket in order to control it: (a) thrust vectoring of the engine and (b) aerodynamic fins. While the aerodynamic fins produce control forces and moments like an aircraft only when the vehicle is moving above a certain minimum inside the atmosphere, the method of thrust vectoring can be used even in the space. In this sense, a rocket is a hybrid of an aircraft and a spacecraft.

A typical launch vehicle has about 90 % of its total mass as propellants, which leaves only about 10 % for the structure, payload, engines, electronics, etc. In the process of maximizing the payload mass for the best launch efficiency, the structural weight must be reduced to within 5 % of the overall weight. This makes for a rather flimsy structure, which cannot take too much of a transverse load before breaking (rather like a can of beverage)! Therefore, it is crucial that both the lift and the sideforce on the launch vehicle should be kept nearly zero. A launch vehicle thus has a *ballistic trajectory*, which is defined as the trajectory with zero-lift and zero sideforce. Since all launch vehicles have an axisymmetric shape (with the exception of the erstwhile *Space Shuttle*) the task of maintaining it in a ballistic trajectory is reduced to keeping both the angle-of-attack and the angle of sideslip zero. In other words, the axis of symmetry must always be aligned with the flight direction. This is accomplished by a dedicated attitude control system. Most of the launch vehicle failures occur when the vehicle cannot be kept aligned with its flight direction due to some reason. The thrust of the rocket comes from the nozzle at the back, while most of its mass is either at the middle, or the front. This makes a rocket statically unstable with about the pitch and yaw axes (see Chap. 3), and controlling the attitude of a rocket is quite like trying to balance a stick on one's hand. Hence, an automatic control system is invariably required to sense any tilt of the vehicle from its instantaneous flight direction, and to immediately apply an opposite torque for correcting the tilt. Such a control system continuously senses the vehicle's rotation from the velocity vector via rate-integrating gyros, and applies a corrective torque by slightly tilting the rocket nozzle in the required direction (called *gimbaling* the engine). Figure 7.4 is a depiction of an attitude control system for a rocket about the pitch axis. The control system consists of a feedback loop between the sensed pitching rotation and the gimbaling actuator (electrohydraulic motor) via a controller, which could be a simple amplifier with a constant gain. The electrical signal to drive the gimbaling actuator is thus made dependent upon the pitch angle measured from the instantaneous velocity vector. As soon as the vehicle has a misalignment with the velocity vector, it produces a small lift (as shown in Fig. 7.4) due to the non-zero angle-of-attack, which is immediately canceled by the normal component of the thrust vector created by gimbaling. Therefore, a sideways motion of the vehicle is prevented and instead a pure pitching moment acts about its center of mass for canceling the pitch rate and restoring the vehicle's alignment with the flight direction. A similar control system ensures a zero sideslip angle by gimbaling the nozzle about the yaw axis (i.e., the axis normal to the pitch axis as well as to the vehicle's axis of symmetry).

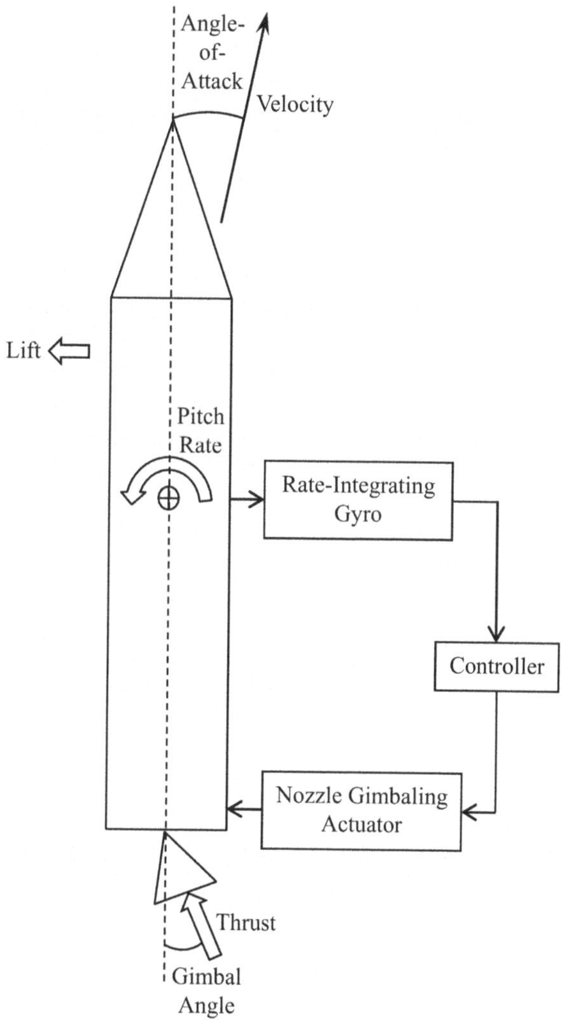

Fig. 7.4 Schematic diagram of the pitch attitude control system of a ballistic rocket. The gimbaling of the nozzle applies a corrective pitching moment in feedback with the sensed pitching rotation supplied by the rate-integrating gyro

The presence of aerodynamic fins located axisymmetrically around the longitudinal axis can generate some restoring pitching and yawing moments (like an aircraft's tail) whenever the angle-of-attack and sideslip, respectively, become non-zero. Furthermore, there is a dynamic effect which comes into play with an increasing speed, and provides resistance with respect to changes in the flight direction. This resistance—explained by Newton's laws—is like the stiffness of a spring, and does not require aerodynamic forces (thus is present even in the space), but only a forward momentum sufficiently large to resist lateral deviations. The natural balancing tendency due to the increased flight speed thereby provides a static stability to the vehicle and is aided by the aerodynamic moments from the fins. Unfortunately, both the help from the fins and the dynamic stiffness provided by a rocket's forward momentum are unavailable when the vehicle has just left its launch pad, because the airspeed at that time is nearly zero. Therefore, a separate mechanism to supply the restoring pitching and yawing moments by gimbaling the engine is invariably required. As the rocket gains speed, its static stability increases due to inertial stiffness effect and the presence of the fins, and after some time in the flight there is no need for the restoring moments to be separately generated by engine gimbaling. This explains why many launch failures happen immediately after launch where the attitude control system may either fail to supply the required balancing moments, or perhaps over-supply them. The need to provide just the right external restoring moments at launch can be understood from our experience with firecracker rockets, where a stick at the back keeps the rocket essentially straight immediately as it is launched. A similar support can be provided by a launch tube or a rail.

Sometimes for stability and control purposes, it may be desired to rotate the rocket about its longitudinal axis. This is called *rolling*, and is most easily carried out by a differential deflection of a pair of the aerodynamic fins located symmetrically about the longitudinal axis. A limited pitch and yaw control can also be provided by a pair of aerodynamic fins when they are longitudinally placed, straddling the center of mass. Figure 7.5 shows a schematic diagram of how an attitude control system based upon aerodynamic fins is implemented on a rocket.

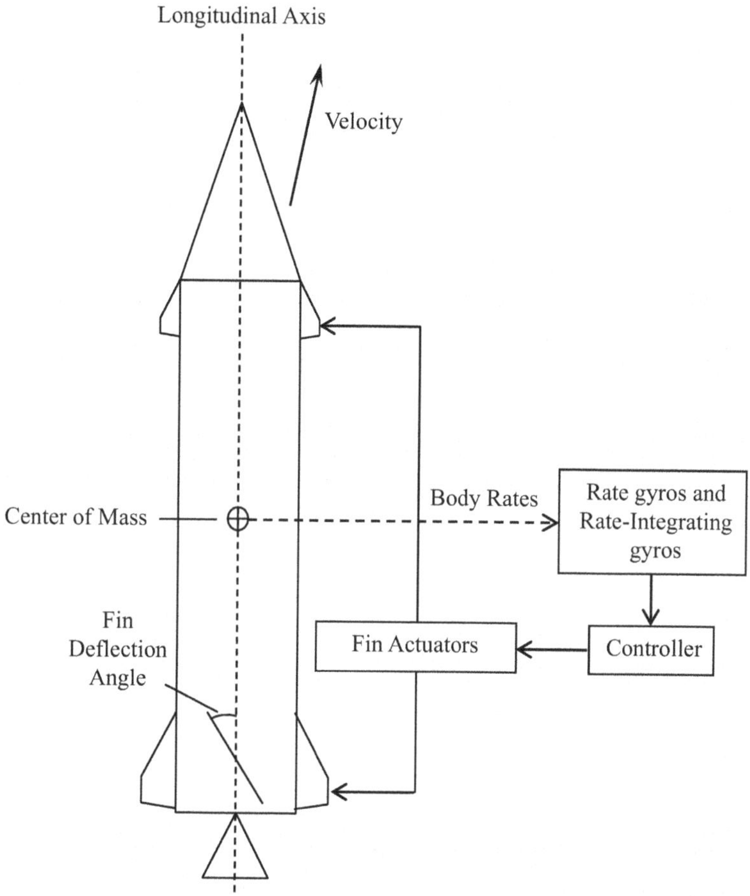

Fig. 7.5 Schematic diagram of the roll attitude control system of a ballistic rocket. A differential deflection of axisymmetrically placed aerodynamic fins at a particular longitudinal position produces the rolling moment, whereas that of fins located at the either side of the center of mass generates pitching and yawing moments. The fin deflections are actuated through a feedback control system, which senses the three body rates via rate and rate-integrating gyros

Appendix: Standard Atmosphere

The *atmosphere* is a thin layer of gases bound to a planet by gravitational attraction. The forces, moments, and heat transfer experienced by an atmospheric flight vehicle depend on the thermodynamic properties of the atmospheric gases, which, in turn, are variables of gravity, planetary rotation, chemical composition, solar radiation, and planetary magnetic field. A model of thermodynamic properties of the atmosphere is necessary for the analysis and design of aerospace vehicles.

At the low altitudes (less than 25 km on the Earth) of flight of most aircraft, the atmosphere can be assumed to be in a thermal equilibrium, with negligible external influences such as electromagnetic disturbances, solar radiation, and chemical reactions. However, this layer of the atmosphere is always undergoing variations in its properties due to winds and the presence of vapors (such as the water on the Earth and the carbon-dioxide on Mars). Such variations are collectively called the *weather*, such as evaporation, condensation, precipitation, lightening, and convective winds. All thermodynamic properties are greatly variable with the altitude, even if the weather phenomena is averaged out over time. A suitable model of the atmospheric considers the vertical variation of the thermodynamic variables, and neglects the temporary effects caused by the weather.

The basic variables representing the thermodynamic state of a gas are its *density*, *temperature*, and *pressure*, which were defined in Chap. 1. The density is expressed in the metric units of kilogram per meter cubed and written as (kg/m^3), and has two major effects. The weight of a given volume of gas is directly proportional to its density, hence a stationary column of the gas would exert a force directly proportional to the density. Secondly, the density determines the inertia of a flowing gas (recall Newton's laws of motion from Chap. 1). A denser packet of gas is accelerated to a smaller speed, when a force is applied on it. While the first effect is useful in deriving the loads on a vehicle due to a stationary gas (called the aerostatic or buoyancy effect), the second feature comes into play whenever we are interested in understanding the aerodynamic loads.

A. Tewari, *Basic Flight Mechanics*, DOI 10.1007/978-3-319-30022-1

A gas consists of a large number of infinitesimal particles (molecules), which are always in random relative motion due to mutual collisions, even though their net motion in any given direction (called the flow) is zero. The *temperature* is a measure of the average *kinetic energy* of a gas particle. A gas at a higher temperature has particles moving at a higher average speed. The standard metric unit of the temperature is Kelvin, written as K.

The *pressure* exerted by a gas on a solid surface is defined as the net rate of change of normal momentum of the gas particles striking per unit area of the surface. The metric units of the pressure are Newton per square meter, written as N/m^2. By Newton's second law, the exchange of normal momentum between the gas particles and the solid surface is responsible for a normal force applied by the gas on the surface. However, it is not necessary that a solid surface be actually present at a point for the pressure to be defined at that point. The pressure can be regarded as the force per unit area that a hypothetical solid surface would encounter, were it present at the given point. By *Pascal's law* (Chap. 1), a gas at rest has the same pressure at all the points. By contrast, a flowing gas has pressure varying from point to point.

A gas flowing past a solid surface also imparts a tangential momentum that is responsible for the force of relative friction per unit area, called the *shear stress*, as discussed in Chap. 2.

While it would be good to have a mathematical model that accounts for the variation of the thermodynamic variables at all the points in the atmosphere and at all times, it is neither practical, nor necessary to do so. An aircraft is not designed to fly only at a particular place and time. Therefore, it is much more practical to take an average of the horizontal and temporal atmospheric variations for deriving a *standard atmosphere*, which contains a model of only the vertical variation of the thermodynamic variables (temperature, pressure, and density). Such a standard atmosphere can serve as a common reference for the calibration of flight instruments, as well as a benchmark for the regulated operation of aircraft.

A standard atmosphere model for the Earth consists of consecutive layers of specified temperature variation with the altitude. The planet is assumed to have an average spherical shape (resulting in a uniform gravity field), and its surface is termed the *mean sea level*. The molecular weight of the atmospheric gases is also variable with the altitude, especially at the high altitudes where the gas is rarefied. Hence, a molecular temperature is defined as the product of the temperature with the ratio of the actual molecular weight to that at mean sea level. The prevailing thermal and aerostatic equilibrium are represented by layers with a linear variation of the molecular temperature with the altitude, whereas the non-equilibrium phenomena at high altitudes generally requires a nonlinear model of temperature variation. While there are several standard atmospheric models for Earth, we shall focus on the *US Standard Atmosphere* of 1976.

There are certain well defined atmospheric strata in the *US Standard Atmosphere* of 1976, as depicted in Fig. A.1. The lowest layer called the *troposphere* extends from the standard sea level up to the 11 km altitude, which is termed the *tropopause*. The temperature decreases linearly with the altitude in the troposphere. The next

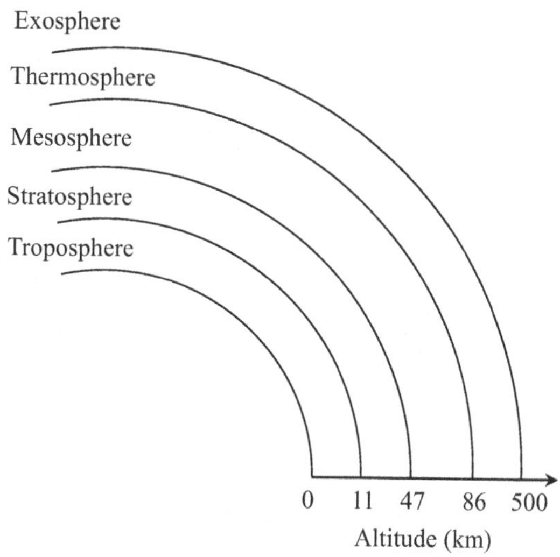

Fig. A.1 Earth's standard atmospheric model shows five strata ranging in altitude from 0 (standard sea level) to higher than 500 km (exosphere)

higher layer, ranging in altitude from 11 to 47 km, is called the *stratosphere*, and consists of three layers with constant (*isothermal*) and linearly increasing temperature at different rates, respectively. The operation of all aircraft is limited to the troposphere and the stratosphere. Immediately above the stratosphere lies the *mesosphere*, which extends up to 86 km and has an isothermal layer, along with two consecutive layers with linearly decreasing temperature. With the mesosphere, the assumption of thermal equilibrium ends, and the next stratum ranging in the altitude from in the range 86 to 500 km, is called the *thermosphere*, which experiences a nonlinear decrease in the molecular weight with altitude due to thermal non-equilibrium and associated chemical reactions. The thermodynamic properties of thermosphere are strong functions of solar radiation, especially such periodic phenomena as the sun spot activity. Beyond the thermosphere lies the *exosphere*, which is an indefinite region dominated by electromagnetic effects of an ionized gas due to the interaction between solar wind and the earth's magnetic field. The exospheric temperature is usually considered a constant with the altitude.

The variation of the temperature with the altitude up to and including mesosphere is plotted in Fig. A.2, whereas the plots of density and pressure are shown in Fig. A.3. At the small altitudes of these strata, the variation in the molecular weight with altitude can be neglected, thereby approximating the molecular temperature by temperature. Note the various linear regions in Fig. A.2, and the rapid decay of both density and pressure with altitude in Fig. A.3. Above an altitude of approximately 50 km, the density and pressure become almost negligible. The atmosphere above 50 km is thus considered to be rarefied.

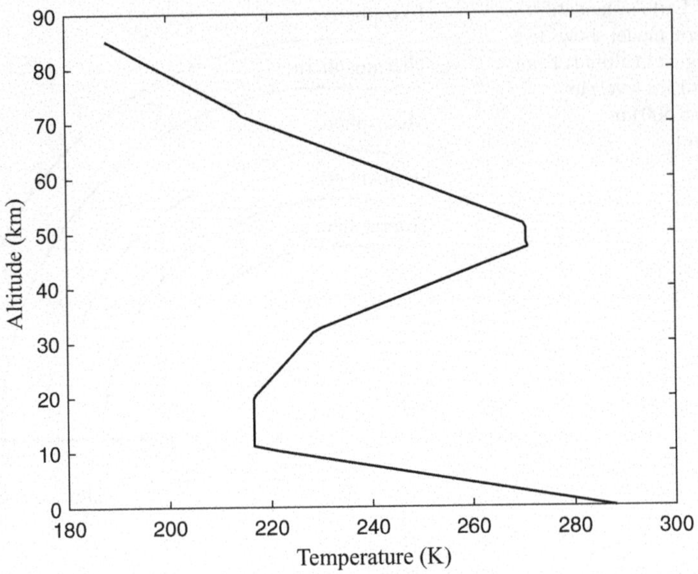

Fig. A.2 Variation of the standard atmospheric temperature in Kelvin (K) with the altitude (km) in the *US Standard Atmosphere* of 1976 for the first three strata (troposphere, stratosphere, and mesosphere)

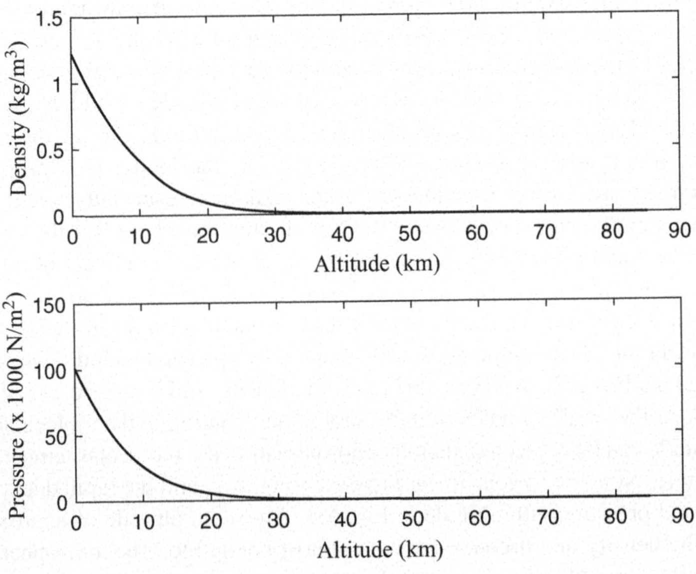

Fig. A.3 Variation of density (kg/m^3) and pressure (N/m^2) with the altitude (km) in the *US Standard Atmosphere* of 1976 for the first three strata (troposphere, stratosphere, and mesosphere)

The manufacturer's authorised representative in the EU is Springer Nature Customer Service Centre GmbH, Europaplatz 3, 69115 Heidelberg, Germany. If you have any concerns regarding our products, please contact ProductSafety@springernature.com

Printed and bound by CPI Group (UK) Ltd, Croydon, CR0 4YY
22/09/2026
02221059-0018